The End *of* Population Growth

Reaching Humankind's Planetary Limits

UGO BARDI

This book is dedicated to my two children, Donata and Francesco; to my four grandchildren, Aurora, Beatrice, Liliana, and Nathan; and to my eight great-grandchildren, who still have to be born.

Disclaimer: This book reports personal opinions of the author, and nothing in it should be seen as necessarily approved or supported by the institutions and associations of which the author is or has been a member.

Images are reproduced under fair use for purposes of criticism, commentary, and scholarly analysis. All copyrighted materials are the property of their respective owners and are used solely for educational purposes. Cover image by Tanja Prokop.

Published by the Club of Rome, 2026
Lagerhausstrasse 9, 8400 Winterthur, Switzerland
ISBN 978-3-033-11888-1
ISBN 978-3-033-11889-8 (epub)

Presentation by Silvia Zimmermann del Castillo

Co-President of the Club of Rome

We live in an era of profound transitions. After centuries of explosive growth, our societies now face the unsettling threshold of population decline. With this book, Ugo Bardi invites us to step into a turning point that is no longer a distant scenario, but a reality that grows stronger each year: the imminent decline of the world's population.

In his earlier work, The Limits to Growth Revisited (2011), Bardi reflected on how key variables such as population, industrialization, food production, pollution, and resource depletion pointed toward a potential collapse in the twenty-first century. Today, this warning takes on a new dimension: it is not only about halting exponential growth, but about understanding the impact of its reversal.

With rigor and sensitivity, the author guides us through system dynamics models, illustrating how forces that once pushed humanity upward are now beginning to shift downward. We see historical patterns repeating: dense societies losing vitality, productive systems slowing down, communities aging, and entire regions reshaping their demographic profiles.

In this context, population decline is not merely a one-sided threat but a critical opportunity. It may force the reinvention of our social, economic, and cultural structures. It will be a rethinking of priorities in which sustainability, equity, and long-term knowledge can return to the heart of the debate.

This book is, therefore, a calm yet uncompromising reflection. Ugo Bardi combines scientific analysis with a humanistic warning, understanding that this new phase is essential if we are to navigate it with wisdom and resilience. It is an invitation to accompany this transition, to look beyond the mirage of unlimited growth, and to envision new ways of living together – both with our planet and with one another.

Presentation by Paul Shrivastava

Co-President of the Club of Rome

From the beginning of its foundation, the Club of Rome's core 'problematique' was understanding the conditions for the long-term survival of humans on Earth. At the center of this challenge lies the size of the human population. The Limits to Growth (1972) report pointed to a tension between a growing and resource-consuming human population and Earth's capacity to support it. This tension remains to this day, after many decades of studying population growth and dynamics. Now we have arrived close to the plateau of population growth. Hence, it is very timely and appropriate to directly confront the phenomenon of the end of population growth.

Ugo Bardi's *The End of Population Growth* is a truly transdisciplinary exploration of a fundamental challenge of the Anthropocene era - the balanced growth of the human population within planetary boundaries. The issue of population growth has been on the minds of policymakers for over three centuries. Both overpopulation and underpopulation are problematic. How do we arrive at the right-sized population on Earth? This is a question fraught with deeper questions and underlying assumptions.

This book traces an evolutionary history of the human population and the factors that have limited population growth in different periods and places. By the mid-20th century, many of the factors limiting population growth were made unimportant by technological progress, and the world experienced an unexpected population explosion, along with numerous attempts at population control. The book offers very interesting case studies of our worldwide population experiences, trends, rapid decline/growth and stability, in the wide ranging narratives that include the fall of the Roman Empire, the great famine and Black Death in Europe, the colonizing of Greenland, the Seneca collapse of Ireland population, Japan's population stabilization, government birth control in China, depopulation in Soviet Union, and explosive growth in Ethiopia, all offer vital lessons on managing the size of populations.

Written in a rigorous and yet entertaining style, the book can help us think deeply about the enduring problem of balancing population and Earth's capacities. We take heart in Bardi's cautious yet hopeful prospects for solving future population challenges. Population decline is not a bad thing in itself, but it may be "too much of a good thing". We still need to balance a complex set of factors, including socio-economic factors and pollution, stresses of urbanization, and avoiding famines and wars. We hope you will find this well-researched book both informative and stimulating.

Foreword

Carlos Alvarez Pereira, Secretary General, The Club of Rome

Let me start by saying something about the author of this book, Professor Ugo Bardi. I have been privileged enough to work with him on several occasions and most notably as co-editor of the book Limits and Beyond, published in 2022. Anyone who knows him is impressed by his unique combination of scientific knowledge, wide culture about all ages of life on Earth, and an unabated sense of humor, which sometimes leads him to craft creative stories with a touch of benevolent sarcasm. You will for sure find some of that wit in the style of this book and, most importantly, the work done shows Ugo Bardi´s infinite curiosity and passion for better understanding complex systems, such as human societies. In that sense, he reminds me of his outstanding compatriot and founder of The Club of Rome, Aurelio Peccei (1908–1984), who never stopped asking questions of relevance to the future of humanity.

This book does exactly that by addressing one of the big questions, which we can bluntly formulate as follows: Are there or will there be too many humans on our planet, considering their impact on the ecosystem of a finite Earth? This question has been haunting us, at the very least, since the times of Thomas Malthus. Hasty responses have often been used to discredit the possibility of ecological humanism, a line of thought and action that could combine the proud care for humanity as a whole and for every one of its members with the cautious care for the health of the ecosystem of which we are part.

As Professor Bardi shows in this book, an informed response to that fundamental question requires considering a multiplicity of factors that evolve over time, or in other terms, adopting the perspective of dynamical systems and the tools used to produce rigorous and relevant knowledge about such systems. This leads to a nuanced assessment of the evolution of the size and age composition of human populations. While growth seems to be a universal pattern, it is actually not, and, most importantly, the size of the global population could stabilize or even decline in the future. To what extent this will happen is impossible to predict, but this book provides the foundational knowledge that is needed for everyone to have a better understanding of how populations evolve and what the consequences are. Against all odds, betting that we have to fight each other to preserve our individual future, there is growing evidence that everyone on Earth could have a very decent life within the limits of the ecosystem. For that, we need to (re-)learn that our health and well-being depend on the fate of everyone else, human or not, and not so much on our level of consumption, as this book also explains. The Club of Rome has been saying for a long time that equity is the essential dimension to consider if we care for both humanity and the planet, or, to be more precise, for humanity as part of a larger whole that we call the planet.

To finalize, there are two aspects that I would like to emphasize about this outstanding work that you are about to read, summarizing my personal interpretation of the book. First, it confirms not only that we have the appropriate tools to understand and observe population dynamics, but that this knowledge can be widely shared with the public at large, so much so that it could help to learn how to navigate together the complex challenges of our coming adventures as human collectives, at multiple levels. Second, and most importantly, it also gives confidence that such collective learning can lead to a reconciliation of humanity with itself and the Earth. Then, we will no longer ask if we are too many or who is in excess, since the existence of each and every one of us will contribute to the flourishing of all and to the continuation of the amazing history of Life on Earth.

Table of Contents

1. Summary

This book is an ideal continuation of the first report to the Club of Rome, *The Limits to Growth* of 1972. It was one of the first quantitative studies that examined global population trends embedded in the whole economic system and the ecosystem. When it was presented, it was misunderstood and widely rejected, but, more than half a century afterward, we see how well it described the trajectory that humankind would follow in the subsequent decades.

Today, we are close to the critical moment when conventional economic growth becomes impossible on a finite planet, constrained by two parallel factors: resource depletion and pollution. These two factors have evolved to forms that were not exactly identifiable at the time of the first dynamic studies, but they are starting to be detectable in our world. The depletion of fossil fuels and other mineral commodities is placing heavy constraints on both industrial and agricultural production. We are not running out of anything yet, but the cost of extraction is increasing, just as the damage that extraction causes to the ecosystem. On the other side, pollution is appearing in more than one form. Chemical pollution is growing in terms of heavy metals, endocrine-disruptors, and other poisoning substances. Climate change can be seen as another form of pollution generated by the excess of CO_2 in the atmosphere.

These constraints are slowing down economic growth and, eventually, the population trend will be affected, too. According to the first version of the *Limits to Growth* model, population decline would start during the first half of the 21st century. Later, similar models based on the same assumptions have confirmed that result. Today, although the world's population is still growing, there are evident signs that we are reaching this turning point: an epochal change that will reverse the trend of growth that had started at the beginning of historical times. We are already seeing the population decline in several industrialized countries, while traditional agriculture-based countries are moving in the same direction, although not yet declining.

The impending global population decline is mainly the result of a global decline in fertility, often attributed to social factors, particularly the trend for women to attain high education levels and engage in professional careers, hence delaying childbearing. However, another explanation is emerging, not as an alternative, but as an additional factor: chemical pollution affecting the human reproductive system and reducing fertility. This factor is often ignored in the debate, but it may turn out to be not just significant, but the most important one.

Governments are reacting to the fertility decline with policies that aim at reversing the trend. The results, so far, have been dismaying. Population decline appears to be deeply embedded in the structure of modern societies in terms of socioeconomic factors and widespread pollution. It is extremely difficult to reverse a trend of fertility decline that, by now, has been ongoing for decades almost everywhere.

In some quarters, the fertility decline has been seen as a catastrophe that may even lead to the extinction of the human species. But it is not necessarily a bad thing. A population decline will lead to a lower demand for dwindling natural resources, lower emissions, and, hence, lower pollution. In general, it will lead to a lower degree of stress on a badly overexploited ecosystem, including Earth's climate. Fewer people may also mean fewer wars because there will be no more growing populations in search of living space.

However, population reduction may also bring enormous problems in terms of maintaining the structure of a society that so far has taken economic growth for granted. Shrinking may be so fast that it could bring about the collapse of some expensive structures that have been built over centuries, such as public education systems, public health care, and scientific research. During the transition phase, society will necessarily consist of an increasing fraction of elderly people, who may become a heavy burden for the young. In addition, the growing population in some African countries may lead to a global imbalance, with a trend of emigration toward regions whose population is declining. That may also lead to social and military conflicts.

This book is an attempt to evaluate current trends and project them into the future on the basis of existing data and models. You will find in it a broad sweep of the trends and problems of the human population. It discusses the subject on the basis of more than one viewpoint: biological, historical, economic, and social factors are all considered. In particular, it is based on the previous experience of the Club of Rome examining global problems from a systems dynamics perspective, on the basis of models, but also in terms of qualitative interpretations. It is the approach that was used more than 50 years ago for the *Limits to Growth* report, and that led to fundamental insights into the trends of human civilization.

No book can predict the future. Things change fast in our world, and one day's assessment may quickly lead to a different one the day after. The result is that our modern, sophisticated mathematical models may turn out to be no better than what the Pythoness of the Oracle of Delphi could do in her time, and one may argue that the more complicated a model is, the less reliable it will turn out to be. But what models can do is tell us what may happen, given some initial assumptions. And that can surely help us to be better prepared for a future that looks uncertain, difficult, and fraught with the risk of more conflicts.

2. Introduction

"Like the generations of leaves, the lives of mortal men. Now the wind scatters the old leaves across the earth, now the living timber bursts with the new buds and spring comes round again. And so with men: as one generation comes to life, another dies away." The Iliad, Book 6

Figure 1 - Adam and Eve by Cranach the Elder (probably 1515). Wikimedia, Public Domain.

Today, those who discuss human population trends tend to split roughly into two groups: the catastrophists and the cornucopians. The former group views population growth as a catastrophe that generates wars, famines, and other ills. The latter view it as a blessing that leads to continued progress and economic growth. Remarkably, both catastrophists and cornucopians tend to see the situation in similar terms; that is, they believe that population growth is the normal state of things, an unavoidable phenomenon that should be either stopped or encouraged.

Indeed, humankind has been continuously growing during historical times, reaching today a level of 8.2 billion individuals. Some recent research even indicates that rural populations may have been systematically undercounted and that the actual human population may have passed the 10 billion mark.[1] In a sense, these numbers are a sign of success for human adaptability and technological prowess, but, at the same time, the impact of humankind on Earth's ecosystem is becoming unsustainable. Humans are by far the most numerous large mammals on the planet. The total mammal biomass is overwhelmingly dominated by livestock (≈630 million tons, Mt) and humans (≈390 Mt), with only 20 Mt remaining for all terrestrial wild mammals.[2] To give an idea of the imbalance, consider that for each human being weighing approximately 70 kg, there are only 3 kg of wild mammal biomass. The huge number of humans is stressing all of Earth's ecosystems: the concept of "planetary boundaries" was introduced in 2009 [3], in terms of six related elements, later expanded to a total of nine,[4] which include not just climate change, but synthetic chemicals, plastics, ozone, ocean acidification, atmospheric aerosol, and much more.

As both the population and the economy are expected to keep growing in the future, these problems are destined to become gradually more serious. But is growth really destined to continue? On this point, we may be facing a radical change in paradigm that may make both the catastrophist and the cornucopian views obsolete. Human fertility is declining everywhere; it is called the "Demographic Transition." Fewer children are born every year, and many couples are childless. The global population is still weakly growing, but as the people born during the "baby boom" of the mid-20th century (the "boomers") march toward the Elysian fields, their loss will not be replaced by the younger generations. The trend is going to be reversed, and the global population is going to start declining.

Scientists working in the field of demography collect data about the human population for the past and extrapolate the trends to the future using mathematical models. These models began forecasting an impending global population decline decades ago, when the United Nations started to use statistical demographic models. Their recent forecasts typically conclude that humankind will reach a maximum level between 9 and 11 billion people around 2080, and then start a slow decline or perhaps reach a steady level.[5] Other population forecasts are based on similar methods, such as the one by IIASA[6] or the Shared Socioeconomic Pathways (SSP) models.[7] They all basically confirm the UN results, although a 2020 demographic study by the University of Washington found the peak occurring somewhat earlier, in 2064, at 9.7 billion.[8]

In addition to statistical models, there exist methods that explicitly take into account factors such as climate change, resource depletion, pollution, and other external constraints that may impact population by affecting natality and mortality. They are called "system dynamics" models, and they started being used for demographic studies in the 1960s, when digital computers became widely available. These models tend to see the start of the decline much earlier than the standard demographic models. The first study using these methods was performed by Jay Forrester in 1971[9], and the model generated the global population peak around 2020-2030. A more refined model based on the same methods was at the basis of the famous *The Limits to Growth* 1972 study.[10] The model generated the start of the population decline no later than 2050 in the "base case" scenario, the one that assumed unchanged economic trends and the best available data at the time. It was a first attempt in a field that, at the time, was still in its infancy. But later studies using the same methods confirmed these early results, almost always generating the start of the global population decline before 2050, sometimes as early as the 2020s.

In most cases, the models produced a symmetric bell-shaped population curve, but in some cases, that decline was found to be faster than growth. It is a trend I called the "Seneca Effect" or "Seneca Cliff" [11] from a sentence written by the Roman Philosopher Lucius Annaeus Seneca. The Seneca Cliff is the typical result of a combination of factors that, in complex systems, tend to reinforce each other, generating a rapid crash of the whole system. For instance, a major war, a financial collapse, and the collapse of some critical sectors of the planetary ecosystem might occur together and reinforce each other. That would generate a rapid and uncontrollable collapse of the economic system, with a consequent rapid decline of the human population caused by the impossibility of providing the resources necessary to keep a large number of people alive.

These global trends refer to a global average and don't imply that the situation is the same everywhere. Today, a significant fraction of the world's population, about 30%, lives in countries where the native population is declining, mostly in Eurasia and North America. On the other side of the trend, some 15% of the world's population lives in regions in Sub-Saharan Africa, where the growth rate has been slowing down but still shows a robust upward trend. In the rest of the world, most people live in regions that are still weakly growing, but where the declining natality indicates that they will soon start their population decline. That's the case, for instance, of India.

These declining trends are not the result of the factors that Thomas Malthus proposed in the late 18th century: famines, wars, and plagues. The countries that are declining faster are those that enjoy the best lifestyle, an abundance of food, good health care, and no evidence of large-scale epidemic diseases. Malthus thought that population would stop growing because of an increase in the death rate. Instead, what we are seeing is a decline in the birth rate. That is a slower and softer trend which, nevertheless, is causing the same effect. If births cannot compensate for the unavoidable demise of old people, the population will decline.

In a stable population, the total number of children per woman over her lifetime (total fertility rate, or TFR) should be about 2.1. If it is less than that, the population will decline, but the start of the decline need not be immediate. Even after fertility rates drop below the replacement level, a population with a broad base of children and young adults will continue to produce more births than deaths, a trend that may last decades. For instance, India's Total Fertility Rate (TFR) has fallen to 1.9 in 2023. Still, India's population growth continues due to demographic momentum from a large young population, particularly concentrated in northern states, while southern states are already below the replacement level. But, in the long run, if the TFR stays below the replacement rate, the population must necessarily start declining.

Trends in different world regions don't just depend on local demographic factors, but also on the balance between immigration and emigration. In recent years, immigration has compensated for the declining native population in several Western countries. A good example is Italy, whose total fertility rate (TFR) went below the replacement rate in 1975 and has been declining ever since. It is now around 1.2, one of the lowest in the world. Nevertheless, Italy's population has remained approximately constant, and even grown to over 60 million, five million more than it was in 1975, as a result of immigration. A different case is that of Ukraine, where not only the fertility rate went below the replacement rate in the 1970s, but also emigration further reduced the population, especially in recent times as a result of the ongoing war. As a result, Ukraine's population fell from a peak of about 52 million to the current level of about 37 million.

The global fertility decline that started in the second half of the 20th century came as a surprise. Nobody had predicted it in the 1960s and 1970s, when fears of overpopulation were common. Right now, we are switching to the opposite perception: it is depopulation that is starting to be feared. But what's causing this new trend? The most common explanation for the fertility decline is based on social and economic factors. Women find it more and more difficult to reconcile having children with their education and career, and hence tend to delay childbearing and reduce the number of their offspring. The impact of education on fertility extends beyond delaying childbirth, as it provides alternative life trajectories beyond motherhood. In addition, raising children to maturity is becoming more and more expensive, and many families cannot possibly afford to have as many as their parents and grandparents did.

At the same time, there is clear evidence that chemical pollution is wreaking havoc on the human reproductive system. Among other factors, the average sperm count of human males (and not only human ones) is declining everywhere at rates that could make most males unable to reproduce by the end of the century, or perhaps even before that date.[12] Another factor that could play a role is the stress correlated to urbanization,[13] also negatively affecting the human reproductive system.

Whatever the cause of the fertility decline, it is not necessarily a bad thing, at least if it doesn't generate a drastic population collapse. A relatively gentle decline will ease the pressure on dwindling resources and a heavily strained ecosystem, including our overheated atmosphere. Also, a declining population will completely change the perception of competition among states. The *Lebensraum* (vital space) concept, popular in Nazi times, was the cause of innumerable horrors. Fortunately, the declining population trends may forever consign it to the dustbin of history, where it deserves to stay. Not that people won't be able to find excuses to keep killing each other, but at least it may become more difficult for dictators to convince people that it is a good idea to exterminate their neighbors in order to gain space and resources for themselves.

Nevertheless, a population decline also brings problems that might inflate to gigantic proportions. One is the growing imbalance among different regions of the world. If we extrapolate the current trends, we see that the growth of the sub-Saharan African countries could bring them to reach about 40% of the world's population by the end of the 21st century, while the Northern countries could shrink to well under 20%. If that's the future, we could see a reversal of the global colonization phase of the past few centuries that saw Northern countries, and Europe in particular, spill their excess population over the Southern countries. If the numeric advantage goes now to the

Southern countries, it is their growing population that could spill into the emptying North. It is already happening, and we all know the political and social problems that this is generating. Note also that the African countries are those destined to be hit the hardest by global warming, with the possibility of generating uncontrolled and uncontrollable migrations toward cooler regions in the North. There are many ways in which this situation could lead to wars and other disasters. Nevertheless, the South is following the same population trajectory as the North. Eventually, the imbalance is destined to disappear, although it will take time to balance the worldwide population trends.

Unfortunately, the demographic decline could be so rapid that it could be defined as catastrophic, causing the collapse of the social and industrial infrastructures that enable our social system to function. Among many possible disasters, we would face the loss of the results of centuries of technological and scientific progress while society slides toward a simpler economy similar to that of medieval times. Even extinction cannot be completely ruled out as the short-term destiny of humankind. Clearly, we are entering uncharted territory, but we don't need to be pessimistic. Our civilization is complex, resourceful, and technologically advanced. If we avoid hasty and perhaps violent reactions to a trend that we can manage, but not reverse, the decline will not be a disaster. But it will be if we squander humankind's remaining resources in new wars and conflicts, and a Malthusian-style catastrophe generated by famines and plagues is still perfectly possible. As always, we are moving toward the future blinded and in darkness. The future will not be what we expected it to be, but that is the way the future always is.

3. The Great Cycle of the Human Population

"There are one hundred and ninety-three living species of monkeys and apes. One hundred and ninety-two of them are covered with hair. The exception is a naked ape self-named Homo sapiens.

Desmond Morris, The Naked Ape, *1967*

Figure 2 - One of the earliest images of a human being: the "Venus of Brassempouy." This remote ancestor of ours lived probably some 25000 years ago. Photograph by Jean-Gilles Berizzi. Wikimedia, Public Domain

The appearance of the creatures we call hominins, the ancestors of modern human beings, marked a momentous change in Earth's history that tilted the biosphere's balance in directions that had never been seen before. It was the result of a combination of factors that turned a species of forest primates into successful savanna hunters, among other things, having them shed their body hair and become "naked apes," according to a definition that Desmond Morris used in his well-known book with the same title, *The Naked Ape* (1967). These creatures, called Homo sapiens, have reached a population so numerous that it greatly surpasses that of any other large animal on Earth in terms of numbers and mass. This huge population cannot continue growing forever, but to understand what the future has in store, we need to understand the past. It all started a few million years ago, and the story is still unfolding.

The Hunting Ape

Humans are the only species on Earth able to manage symbolic speech, a capability that gives them immense power to create and destroy at their will. This capability is related to the special features of their brain, which is not just a larger version of the typical mammalian brain, but also a differently structured instrument endowed with a high neuron density. [14] Humans even have a special gene that generates this tight neuronal packing. [15] Only a few other mammal and bird species have this kind of sophisticated brain, a novel characteristic of the biosphere. It appeared in relatively recent times, no more than a few million years ago. It has been speculated that this could be due to the high atmospheric O_2/CO_2 ratio of recent geological times, which reduced the metabolic cost of oxygenating densely packed brains.[16]

The high computing power of the brain is a characteristic not just of modern humans but also of their recent ancestors, the hominins (or *Hominini*), a "tribe" within the family of Hominidae, which includes the great apes. The first hominins appeared in central Africa approximately four million years ago, in the form of Australopithecines. The most famous of them was "Lucy," the name given to a skeleton found in 1974 in Ethiopia.

The Australopithecines' brains were no more than 35% of the volume of a modern human's brain. We have no evidence that they used tools or fire, nor that they could hunt large prey; they seem to have been vegetarian. They were probably also hairy, like modern apes. But the australopithecines already had some of the typical characteristics of the hominins, the main one being that of walking erect most of the time. They had left the forest environment of the great apes to try their luck in grasslands and savannas. There, bipedal walking was an advantage in terms of efficiency and also allowed them a large field of view, surely valuable in an area populated by much larger carnivores. It is likely that they already had sophisticated social skills, allowing them to forage in groups and to help each other.

Hominins lived in central Africa for at least four million years, improving and refining their skills and capabilities. They lost their body hair,[17] and hence became able to cool their bodies by sweating at rates much larger than those of other mammals. Their social abilities allowed them to replenish their bodies with water with the help of other group members and perhaps of one of the earliest technological tools in human history, the water gourd. In this way, humans could develop an endurance that no other large land creature on Earth could match.

With the appearance of *Homo habilis* or *Homo erectus*, starting around 1.5-2 million years ago, we have the first evidence of stone tools and the capability of controlling fire. These skills allowed hominins to attack large animals and to cook meat before eating it. That removed the need for the heavy jaws and cumbersome digestive apparatus characteristic of other large carnivores. Richard Wrangham described how cooking made us what we are in his book *Catching Fire* (2009), and we might take from there a possible definition of human beings as the "Cooking Apes."

The development of these skills was a gradual process. It led to an evolutionary "arms race" with the African herbivores, which had to match the abilities of the new carnivores. The large ears of African elephants evolved in parallel with hominins, and we can see them as a device to cool elephants' bodies and enable them to endure the effort of long chases.

The arms race also involved competition with other carnivores, and not all of them survived it. The African sabertooth cats, such as Dinofelis, disappeared from sub-Saharan Africa around 1.7 to 1.5 million years ago, almost certainly as the result of competition with hominins. It is the harsh biological law called "Gause's law" that says that only one species can occupy a single ecological niche. On the other side of the Atlantic Ocean, the iconic "Sabertooth Tiger" (Smilodon) lingered on up to relatively recent times, about 10,000 years ago. Still, it didn't survive the arrival of modern humans on the American continent.

The physical characteristics and the skills of hominins were optimized to make them effective hunters in the African savannas. But, as often happens in an evolutionary game, a feature developed for a specific purpose turns out to open new possibilities of adaptation in another environment. This process was termed "exaptation" by Stephen Jay Gould.[18] It was exaptation that allowed the African hominins to start a venture that would lead them to boldly go where no primate had gone before: to the cold boreal lands of Eurasia.

Cold regions pose a peculiar survival problem to their fauna: how to survive in winter when plants go dormant and produce no food? Small creatures can burrow and then hibernate in relative safety. Large animals cannot burrow, and hibernating in the open would be dangerous for them if discovered and attacked by carnivores. For them, a better strategy is to spend the winter on their legs. So, they accumulate fat in their bodies during the good times, slowly consuming it throughout the winter. Large herbivores also have a digestive system that allows them to eat the poorly digestible food still available in winter, such as tree bark. Carnivores can remain active in winter, finding a convenient form of "meat on legs" in the large herbivores.

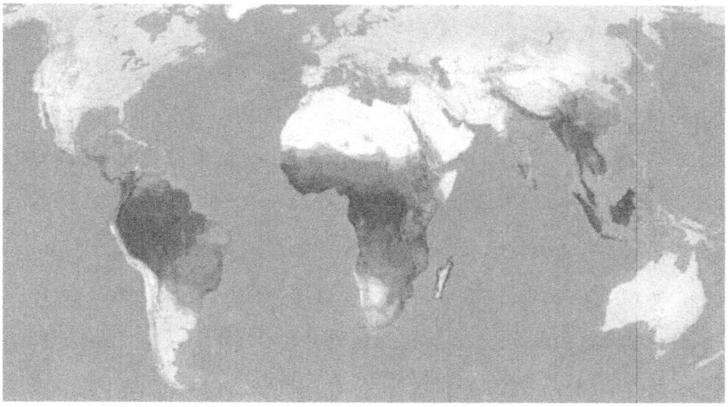

Figure 3- The current range of non-human primates (darker areas). For most of their existence, some 60-70 million years, they remained tropical animals, never able to expand into cold boreal regions. Wikipedia, Creative Commons.

But primates can't follow any of those strategies. They are relatively small creatures adapted to an arboreal lifestyle and a diet of fruit, seeds, and nuts. A hypothetical "boreal monkey" would find no food in winter in the icy forests of the North, nor in the snowy steppes. So, today, the northernmost living monkeys are some Japanese macaques, known to bathe in hot springs to keep themselves warm in winter.

Primates needed to develop new and different characteristics if they were to attempt to venture into the cold Northern regions. Hominins happened to be able to do just that by means of the tools and the skills they had developed in the hot climate of Africa. High endurance, group hunting, and stone weapons allowed them to hunt large prey, such as woolly mammoths and bison, and provided abundant meat available throughout winter. Their social habits allowed them to stock and defend a supply of food that wouldn't spoil because of the cold. Fire was useful not just to cook meat but also to keep warm, and tanned pelts made using stone tools could also help. With these skills, surviving the harsh winters of ice ages was possible, even though some recent data indicate that Neanderthals could also hibernate, a skill probably needed in the extreme cold of ice ages. [19]

Several waves of hominins emerged out of Africa and moved into Eurasia. The first was probably *Homo Erectus* who reached Eurasia about two million years ago. About 400,000 years ago, the Neanderthals (*Homo Neanderthalensis*) appeared in Europe and in the Middle East. The Denisovans (*Homo Denisova*) were another hominin species living in Asia at the same time. These Eurasian populations always remained small: the Neanderthals never were more than about 70,000 individuals. Their impact on the ecosystem was relatively small, although not negligible. For instance, they used fire to control forest vegetation density. [20]

Everything changed when our direct ancestors, the Sapiens, moved out of Africa and into Eurasia, about 60,000-70,000 years ago. Their toolkit was not fundamentally different from that of the earlier Eurasian hominins. But, evidently, the Sapiens had a larger range of technological and social skills, and that gave them a competitive edge in the Eurasian environment. The Sapiens wave into Eurasia was the equivalent of an invasion of aliens in modern science fiction movies. They crossed plains and mountains and even oceans, marching on the land bridges created when the sea level became low enough during the coldest periods of ice ages. A journey that was concluded when the last uninhabited islands of the Pacific Ocean were settled by Polynesian people, less than one thousand years ago.

We have no evidence of direct conflicts between the Sapiens and the other hominins living in Eurasia. On the contrary, the encounters must have been friendly on more than one occasion since modern Europeans still carry some Neanderthal genes. But Gause's harsh law condemned the less efficient species, the Neanderthals, who disappeared from Europe some 50,000 years ago. The Denisovans may have disappeared at about the same time, but they, too, left traces of their existence in the genome of modern Asians. [21]

The human wave that swept the continents didn't just replace the existing hominins. It went together with a wave of extinctions of large herbivores (the "megafauna"). Mammoths, woolly rhinoceroses, giant sloths, and even horses (in North America) disappeared, engulfed by the great wave that swept first Eurasia, then the Americas, and finally Australia and Oceania. The same fate awaited non-human carnivores in competition with humans, including the European Cave Lion *(Panthera spelaea)*, which most likely never lived in caves. We have splendid images of these lions painted by our ancestors in the Chauvet cave in France. They may have admired those magnificent beasts; nevertheless, they drove them to extinction.

This wave of extinctions may have been the first case of humans overshooting their resources and finding themselves in a condition of overpopulation. We don't have sufficient data to determine if that was the case. Still, without their traditional large prey, our hunting ancestors had to adapt to smaller prey. Perhaps as a result of this ancient overpopulation crisis, humans developed a new technology that was going to engulf Earth: agriculture.

The Agricultural Age

With the end of the last Ice Age, about 12,000 years ago, warming generated a mild climate and reliable water sources from rivers such as the Nile, the Tigris, and the Euphrates. Early farmers cut down forests to leave space for agriculture and began to domesticate plants like wheat and barley. The earliest evidence of human agriculture dates back to around 9500 BCE, when the Natufian population in Egypt first cultivated wild grains.

We don't know whether agricultural practices spread from the Levant to other parts of the world or were independently redeveloped. In any case, by 7000 BCE, farming had reached a significant fraction of Europe, and by 6000 BCE, it appeared in the Indus Valley in South Asia. In China, rice cultivation began around 7000 BCE. In the Americas, maize (corn) was domesticated in Mexico around 6000 BCE.

Agriculture was another case in which tools developed for a specific purpose were found to be helpful for a completely different one. Stone tools were used by humans mostly for hunting, but they turned out to be useful for cutting down trees. Fire could be used to cook meat, but also for clearing forests to create space to spread seeds on the ground. The social skills developed for group hunting turned out to be usable also to manage and defend cultivated fields. Later, tilling, still using stone and wooden tools, was developed to remove unwanted grasses and put the new seeds directly in contact with the fertile soil.

In these early times, farmers relied on rain or other natural water sources for irrigation, but it soon appeared clear that water had to be actively provided to the fields. The ancient Egyptians developed irrigation systems along the Nile River, allowing them to grow crops year-round. The Sumerians in Mesopotamia also built irrigation canals, which significantly increased their agricultural output. Crop rotation was also developed to maintain soil fertility.

Agriculture allowed a much larger population density than anything that could be sustained in earlier times by hunting and gathering. That led to the growth of permanent settlements. With cities came metalworking, trade networks, organized religions, kingdoms, slavery, and much more. The results had profound social and cultural impacts. While hunter-gatherers were jacks-of-all-trades, city dwellers could specialize in different jobs. They became toolmakers, tailors, merchants, scribes, priests, and more. Some could become full-time warriors, and others took up mining, a job that hunter-gatherers rarely could manage at the scale needed to extract metals and smelt them.

Agriculture was a spectacular success, but it had a downside, too. A problem was that the yield depended on the vagaries of weather and other uncontrollable factors, such as insect invasions. That implied the need for long-term storage of the harvest. It could be done with dry grains such as wheat, barley, and others. In optimal conditions, with no mold, no insects, and no mice, grains could theoretically be stored in silos for 2-5 years or even more, even though they would lose part of their nutritional value. The Biblical story of the famine in Egypt is based on historical facts. The inscriptions on the "Famine Stela," located near Aswan, report a seven-year famine during the reign of Pharaoh Djoser. Other ancient inscriptions in Mesopotamia describe famines occurring in these ancient times, and how wise rulers helped their people survive.

Despite grain storage, famines soon became a feature of agricultural societies. It was probably because rulers tended to use the surplus production not so much to store resources for lean years, but to hire mercenaries and embark on wars, or to build pyramids, or for other expensive tasks. So, when hard times came, there would be little or nothing left to feed ordinary people. In practice, the life of farmers all over history seems to have been one of true scarcity, on the edge of starvation. One year of bad weather, or a parasite that attacked the crops, would be sufficient to plunge a population into the abyss of famine. Even in modern times, the traditions and lore of farmers focus on being frugal, always careful to avoid wasting food. An old story still told today in Italy says that when a peasant dies, his soul will be allowed to go to Paradise only after his ghost has found and recovered all the bread crumbs he dropped on the ground in life.

Yield fluctuations were not the only problem with ancient agriculture. Another was the difficulty of providing all the nutrients that are essential for human health. Cultivating a single crop was the way to maximize yield, but a single crop does not provide all the necessary nutrients for human beings. Crop monocultures generated various kinds of sicknesses common in agricultural societies: beriberi, scurvy, pellagra, and others. Some crops are said to be better than others in this sense, and potatoes may be able to provide a nearly complete diet for human beings. But even potatoes lack some essential minerals. So, peasants tended to be shorter and less robust than hunter-gatherers and to live shorter lives.

Finally, the most important problem of agriculture is that it has never been a sustainable activity, at least in the long term. A natural ecosystem is arranged in such a way that it recycles nearly 100% of its nutrients. It has to be so; otherwise, it couldn't exist for a long time. But an ecosystem is not made to sustain large numbers of humans. Humans are always in a hurry; they grow, they produce, they go to war, and they care little about the long-term maintenance of the systems that keep them alive.

The degradation of agricultural soil is a real and pressing problem that has destroyed several ancient agricultural societies. One of the early areas subjected to this phenomenon is the Fertile Crescent, the area we now call the Middle East. Once, it was covered with lush forests, but it was gradually occupied by increasingly dense settlements. People deforested the land to replace trees with barley and grain. They used the water from the rivers that flowed through the region to create a sophisticated irrigation system that compensated for the loss of rain.

Unfortunately, their prowess in irrigation led them to their doom because of the rise of the water table, the boundary between groundwater and the unsaturated zone above it. It is the upper boundary of aquifers: when you dig a well, you have to reach that boundary if you want to pump water out. But, as a result of irrigation, the water table rises, the water evaporates, and salts accumulate in the soil, a process known as salinization, which makes the soil less fertile. Poor drainage worsens the problem.

As early as 2400 BCE, the Sumerians attempted to combat salinization through methods such as leaching and constructing drainage channels. But they never completely understood the reason for their doom, and they failed. They left us an enlightening document with the story of the battle of the Goddess Inanna against the Mountain Ebih.[22] We can see this story as an allegory of a landslide caused by erosion. Other agricultural civilizations later replaced the Sumerian one in the Middle East, but the problem remained. Today, the area is mainly a desert where agriculture is kept alive by artificial irrigation and fertilizers.

It is said that in their march toward the future, people have forests in front of them and deserts behind them. It is a good way to illustrate humankind's path over the past few millennia. Always moving onward, always cutting, seeding, harvesting, and moving on, away from the growing desert. At the beginning, there was plenty of fertile land to expand into, but then the land was gradually filled with farms, and overexploited areas became deserts. Humankind may have faced its limits in food production by traditional methods during the last centuries of the second millennium CE. But the Green Revolution allowed a new leap forward, bringing human numbers to the current levels that had never been seen before. Even these numbers, however, are reaching their limits nowadays.

Protein supply: Hunting, Fishing, and Livestock Raising

The old saying that you can't live by bread alone holds not just in metaphorical terms. Human beings cannot live only on crops without experiencing plenty of health problems, becoming vulnerable to epidemics, and living short lives. Our remote hunter-gatherer ancestors had a varied diet that included fruit and meat, and their health seems to have been excellent. But the situation of peasants was very different. Maximizing the agricultural yield of a region meant getting rid of all the previously existing ecosystems, and that was incompatible with hunting and gathering, which required larger areas to provide the same amounts of food.

Hunting didn't disappear, but it became an activity reserved for nobles. Obviously, nobles could produce only meat for their own consumption, and they had no intention to share it with the commoners. Maybe you remember the story of Geordie, an old Medieval song popularized by Joan Baez in the 1960s. According to the story, Geordie was a young man who "stole sixteen of the king's royal deer," and for this, "he will be hanged in a golden chain." A bit harsh as a punishment, but it was the result of limited resources having to be treasured. At best, the nobles would allow commoners to collect berries and fallen branches for their fireplaces. But not only hunting, but also unauthorized tree cutting was severely punished, even with death. It was the case in Japan during the Edo period.

Fishing as a source of protein was a different story because it didn't involve competition for land with agriculture. It is known that our Stone Age ancestors would consume seafood and fish from rivers and lakes. They could also catch fish from the sea near the shore, the kind we call today demersal. The problem was how to access the more abundant fish resources farther away from the shore, the kind of fish we call pelagic: salmon, tuna, sharks, and many others. Many kinds of marine mammals are also pelagic.

Humans are poor swimmers, and they can catch pelagic fish only by using boats. But building an effective fishing boat requires specialized skills, usually beyond the capabilities of hunter-gatherer societies. The Northern Inuit were an exception, probably because their primary food source, marine mammals, was abundant enough to make the effort of building and maintaining boats worthwhile.

The real problem with fishing is that it is an inherently uncertain activity. Sometimes, it would bring miraculous catches, such as when Christ's disciples caught 153 fish in a single run, as you can read in John 21:1-14. Large yields didn't always need divine intervention, but the other face of variability was that sometimes the boats would return home empty. In winter, bad weather would make fishing by boat impossible for months. As a result, ensuring a constant supply of food was an insurmountable problem for fishermen. Fish spoils easily; it is the result of a number of biochemical factors that make fish tissue more easily attacked by bacteria than meat. Even in relatively cold climates, fish cannot be stored in an edible form for more than a few days without special precautions and technologies that were not available to ancient people. So, a family or tribe that relied fully on fishing couldn't have food for more than a few days without a new catch. If they were lucky, they could make it through the summer, but they would almost certainly starve to death in winter.

Things changed radically with the development of agrarian societies. Cities could provide the resources and the knowledge necessary to build boats, while fishing could provide proteins for farmers and town dwellers. It didn't matter so much if the fish catch was wildly variable. When fish were abundant, farmers could have the proteins they needed; when the catch was scarce, the fishermen could survive on agricultural products. The development of metal money facilitated these exchanges; perhaps money was even created by this need, as Ilaria Perissi and I suggest in our book *The Empty Sea* (2021).

The features of this mixed fishing/farming economy based on currency had the unexpected effect of condemning fishermen to eternal poverty. It was a necessary consequence of market factors that operated in ancient times, just as they do nowadays. The fishermen had no control over the yield of their activity, so when the catch was good, the excess supply lowered the price and reduced their profits. When the catch was bad, prices were high, but they had little or nothing to sell. In other words, fishermen could not accumulate capital. It is for this reason that, in fiction, fishermen are almost always depicted as poor, even desperately poor. Even Jesus Christ's disciples in Palestine, mostly fishermen, were poor people on a low rung of the social scale. We have no evidence from the Gospels that Jesus was a fisherman himself, although the mention that he could walk on water seems to indicate that he engaged in some kind of offshore activities.

Although necessary as a source of proteins, fish was considered a vile kind of food in ancient times, most likely because of the difficulties of conserving it. The nobles shunned it, much preferring the more expensive meat. The commoners, probably, had to consume their fish whenever they had a chance to, and hence to ingest a good fraction of putrefaction products. That was a cause of botulism and other kinds of illness, sometimes generating just stomachaches, but sometimes even death. Not for nothing, the Catholic Church allowed fish as a food that could be eaten during penance days. Sometimes, it must have been a real penance. Of course, our modern views on fish have changed a lot, and now we consider it a fancy kind of food. But that's due to the havoc created by overfishing, which has made good fish rare and expensive.

These problems led to early attempts to find ways to protect fish from putrefaction. In some cases, it could be dried simply by exposing fish to open air, but that was possible only in special climatic conditions, such as those existing in Norway. Try to do that in a Mediterranean climate, and your neighbors will probably run after you with their battle axes. But fish dried in Norway didn't smell foul and could be exported. It also provided a transportable source of food for Vikings for their raiding and commercial expeditions. But fish was not sufficient to feed the ancient Norwegians, who had to cultivate grain and other cereals anyway.

Another early technology for fish preservation was salting. Treating the fish with abundant salt kills the bacteria that spoil it, and that makes it possible to return the fish to an edible form by washing it to remove the salt. The method was already known in ancient times, but the problem was the cost. Before the industrial age, mining salt from land mines required expensive efforts and plenty of manpower. The other way to produce it, extracting it from seawater, involves evaporating large volumes of water, which takes time, energy, and space. As a result, salt was an expensive commodity in ancient times. It is because of this that we still use the term "salary," which derives from a special monetary provision to the ancient Roman Legionaries that was paid in order to enable them to buy salt. Another piece of evidence of the high cost of salt in ancient times is the elaborate salt cellars that artists produced for the rich. The one that Benvenuto Cellini made for the king of France in 1545 is still famous today. Commoners, instead, couldn't afford salt and had to eat their food tasteless. But they were probably hungry enough not to notice.

In time, salt production increased with better technologies and expanded markets. By the 1700s, it became possible to market salted fish as a mass product. Cod was caught and salted in the Atlantic Ocean and then shipped to Southern Europe under the Portuguese name of *bacalhau*. Later, canning and refrigeration technologies changed the whole story. By the 1950s, large fishing vessels could process and freeze fish on board. You surely remember Captain Birdseye promoting frozen "fish sticks" or "fish fingers" in the 1950s and 1960s. It was the start of a fishing revolution that made fish a ubiquitous and safe food.

The very success of modern fishing generated an enormous problem: overfishing. Up until not long ago, the sea was considered an inexhaustible source of food, but that's not true anymore. Already in the mid-19th century, some species of oceanic whales had been hunted to near extinction. Earlier on, the Mediterranean whales had been driven to extinction in Roman times. With time, fishing yields dwindled and forced fishermen to go lower and lower in the food chain, from fish to arthropods, and now to krill, small arthropods that, up to now, were eaten only by whales. Today, you can buy krill as pet food in Europe and as food for humans in the US, China, and Japan.

The loss of valuable large fish species led to the development of fish farming, with high-value fish kept in tanks near the seashore, nourished with low-value fish caught in the sea. It was hailed as a positive innovation, but the impact of fish farms on the seashore ecosystem is enormous, and the fish produced in this way don't have the same nutritional quality as wild fish. Will there be a time when the ocean will become empty of fish, just like some once fertile areas became deserts? We can't rule it out, as discussed in books such as *World Without Fish* (2011) by Max Kurlansky. The depletion of the world's fisheries does not bode well for the future of food supply for humankind.

In parallel with fishing, another source of protein for human beings was developed: animal husbandry. It is a technology that stretches back to the dawn of human civilization, evolving alongside our transition from hunter-gatherers to settled farmers. Although husbandry couldn't produce as much food as agriculture for the same area occupied, it didn't require the large areas needed to hunt wild species. So, husbandry was widely practiced for its capability to produce high-quality proteins in the form of meat and dairy products.

The first tamed animals were likely dogs, descended from wolves, around 15,000 to 30,000 years ago. They weren't livestock, but they could be eaten as a last resort food. The real game-changer came with goats and sheep in the Fertile Crescent (modern-day Middle East) around 9000–8000 BCE. These animals were prized for meat, milk, and hides. Cattle followed around 8000 BCE in the same region, likely domesticated from wild aurochs. Pigs came around the same time in both the Near East and China, thriving on scraps and requiring little space. Chickens can be traced back to Southeast Asia's red junglefowl. They joined the human-controlled fauna by 6000 BCE.

By 4000 BCE, animal husbandry had spread and diversified. In Egypt, cattle were sacred and central to agriculture. In the Eurasian steppes, horses were domesticated around 3500 BCE, not for food, but to pull chariots; horse riding came later. Camels, tamed in Arabia and Central Asia around 3000 BCE, opened desert trade routes. Meanwhile, in the Andes, llamas were domesticated by 4000 BCE for wool, meat, and carrying loads, ideally suited to rugged highlands.

The Bronze and Iron Ages saw new changes. Selective breeding emerged, with bigger cows and woollier sheep. In Mesopotamia, texts from 2000 BCE detail livestock management, such as feeding regimes and disease control. The Romans later developed large-scale farms raising cattle, pigs, and poultry, as documented in works like Varro's De Re Rustica (1st century CE). They also experimented with fish ponds and beekeeping. The Middle Ages kept moving along the same lines. Cheese was also a typical food during the Middle Ages. Since it didn't spoil easily, it had military importance, sustaining troops marching toward distant regions.

The arrival of Europeans in the Americas after 1492 was another push toward unbalancing the local ecosystems. Horses returned to the continent after they had been exterminated by the ancient Americans thousands of years before. This return gave rise to the image of the North American Natives as being expert horse riders. It was a skill that they developed in relatively recent times, but it didn't help them much to avoid being exterminated by the European settlers. Large-scale cattle herding was introduced in North America after the near complete extermination of the wild North American bison.

The problem with animal husbandry was that, unlike fishing, it was in competition for land with agriculture. That often led to violent confrontations. Traditional pastoralist societies could not sustain population densities comparable to those of agricultural societies, but the difference was not so large as with hunter-gatherer societies. Since pastoralists tended to occupy extensive areas of land, the result was that they could often field armies as large, and sometimes larger, than those marshaled by agricultural societies. The lower population density of nomadic pastoralists also made them more resilient and less vulnerable to the typical factor that caused the decline of agricultural societies: the degradation of fertile soil.

The result was a historical trend that has been going on for millennia: the conflict between agricultural and pastoral societies, with the pastoralists often giving the final push to a declining agricultural empire. There are many examples, an early one is that of the Gutian sack of the Akkadian Empire in Southern Mesopotamia around the 22nd century BCE. The myth of Cain and Abel, the two sons of Adam, may have been influenced by these ancient events. In more recent times, we had the Huns, a pastoralist population, nearly destroying the Western Roman Empire during the 5th century CE. In later times, we had the Mongols' conquest of the Southern Song Dynasty in China during the 13th century. The latest version of this ancient historical cycle is that of the conflict between farmers and cowboys in North America during the 19th century. In modern Hollywood movies, we often see peace-loving farmers trying to settle in an area previously occupied by wilder cowboys. If you saw films such as *Shane* (1953) or *The Big Country* (1958), you likely didn't forget the gripping stories of conflict they told. But note that in the story of Cain and Abel, the good guy is the shepherd (Abel), while the bad guy is the farmer (Cain). This is not surprising because the Hebrews were originally pastoralists.

In modern times, animal husbandry can produce much more food than it did in ancient times, and does not require nomadic herders living most of their lives riding horses. But that needs an integrated approach with agriculture that ancient societies couldn't manage. Agriculture and animal husbandry have become a set of complicated technologies supposed to optimize the yield of a farming enterprise. Still, these arrangements often create problems of infectious diseases that affect both the animals and their keepers. Diseases affecting animals are often cured by culling or by massive doses of antibiotics. Neither is the perfect solution, and, as usual, it is the excess of optimization that creates fragile systems that tend to collapse and produce more problems than they solve.

The Green Revolution

The Industrial Revolution started in the 1800s, and it turbocharged all economic activities, including food production. It created the basis for the giant leap forward in human population that's still ongoing today. Before the Industrial Revolution, one of the main limits for all economies was the availability of wood. Would-be empires had to play a challenging game to balance their needs to keep healthy forests to produce wood for ships and charcoal for metallurgy, and the need to deforest the land to produce food for a large population. A mistake in managing these opposite needs could ruin an entire state, not just for the lack of wood for ships, but for the soil erosion that follows deforestation. The mighty Spanish Empire of the 16th century, the one on which "the sun never set," didn't play the game well enough, and it gradually disappeared into the dustbin of history.

Instead, Britain found a trump card to win the game: coal. The "Glorious Revolution" of 1688 was not just about replacing one king with another. It ushered in an economic revolution. In the early 18th century, the process of coking coal was developed, changing the destiny of a small island at the edge of Europe, propelling it to become the center of a vast world empire. Using the large coal reserves of England, the British industry could smelt iron in large amounts while, at the same time, sparing the British forests. Timber could then be used to make ships. With wooden ships and iron weapons, the British could ferry mighty armies overseas, defeat foreign kingdoms, and enslave their populations. Then, the conquered lands could be turned into colonies to produce food for Britain: the British had found a way to turn coal into food. Admittedly, a rather convoluted one, but it worked. It was not yet the modern green revolution, but it allowed the British population to swell way above the limits that their island would allow. Later, other European states did the same by using their coal reserves, but Britain was the first and the most successful at that.

The availability of food and fuel ushered in the Industrial Revolution, which started in England in the 18th century. The consequences soon started to be felt in agriculture. For one thing, the industry could produce agricultural tools at low cost, even though the mechanization of agriculture started only in the 20th century. Instead, already during the 19th century, the growth of the chemical industry had a dramatic impact on agriculture.

For millennia, farmers knew the earth needed feeding. Manure, ash, and crop rotation kept fields fertile, but by the 18th century, Europe's growing cities and shrinking farmland demanded more. In 1804, the Swiss chemist Nicolas-Théodore de Saussure showed that plants needed nitrogen, phosphorus, and potassium—nutrients often scarce in worn soils. Then, there came Justus von Liebig, a German chemist. His 1840 book, *Organic Chemistry in Its Applications to Agriculture and Physiology*, started the landslide. Liebig argued that soil wasn't a passive sponge but a chemical system that was depleted by crops.

Early experiments to return fertility to degraded land were crude but effective. In Britain, in 1842, John Bennet Lawes created "superphosphate," a phosphorus-rich powder created by reacting bones with acid. He patented it, built a factory at Rothamsted, and by 1843 was selling it to farmers. Wheat yields jumped 20–30% where it was used. Phosphorus was a fundamental nutrient for plants, but what really changed the game was nitrogen. All plants need it, but the natural supply depends on the laborious bacterial process of fixing nitrogen, that is, removing it from its unreactive form as an atmospheric gas, and turning it into forms that could be absorbed by plants. For a long time, the only way to provide nitrogen to agriculture was to recycle it from manure. Human manure was a common component of the cycle; up to relatively recent times, it had a commercial value as fertilizer. For a certain period, starting in the early 180s, the discovery of huge amounts of accumulated guano (bird droppings) in Peru's Chincha Islands led to a brisk industrial activity of collecting it and shipping it everywhere. But the poor Peruvian birds couldn't poop fast enough for the world's demand, and the stock was overexploited. By the late 1870s, guano production collapsed. Some guano is still produced in the islands because it is cheap to collect, but it is a small amount, far away from the peaks of earlier times.

The real breakthrough with nitrogen came in 1909 when Fritz Haber, a German chemist, started working on ammonia synthesis. It was a revolutionary idea: extracting nitrogen directly from the air by making it react with hydrogen under conditions of high heat and pressure and turning it into ammonia. The hydrogen, in turn, could come from the high-temperature decomposition of fossil fuels, such as natural gas (methane). Another German chemist, Carl Bosch, scaled the process at the industrial level by 1913, giving birth to the Haber-Bosch process. The initial idea was to turn ammonia into nitrate to manufacture explosives. But with the end of World War I and with Germany having experienced a severe famine caused by the Allied blockade, it was soon realized that nitrates were an excellent fertilizer. Germany's ammonia output reached 200,000 tons per year by 1930. It may have been because of nitrates that Germany didn't suffer a disastrous famine during WWII, despite a new Allied blockade. It was only after the war, in the zones occupied by the Allies, that the Germans started suffering from a lack of food as a result of the draconian control of the food supply by the Allied authorities. Fortunately for them, soon the Allies abandoned their plans of starving to death a large fraction of the German population according to the "Morgenthau Plan," which President Roosevelt had approved in 1944.

The ammonia revolution rapidly spread to other countries. In the United States, Haber-Bosch plants for agriculture started appearing in the 1920s and were fueled by cheap oil. By 1940, U.S.

fertilizer use had tripled from 1900 levels. In the Midwest, corn went from 1.5 tons per hectare in 1920 to five tons by 1970. Colonial powers exported the tech to overseas lands. In India, Punjab saw rice yields climb from 1.5 tons per hectare in 1960 to 4 tons by 1980. The Soviet Union built Haber-Bosch plants in the 1930s, but not fast enough to avoid the *Holodomor,* the great famine that affected mainly Ukraine but also Russia from 1932 to 1933. Japan embraced fertilizers early. By 1900, they were using superphosphate for rice paddies, doubling yields, even though that was not enough to avoid a severe famine in Japan during WWII. This Japanese famine is shown in the movie *The Grave of the Fireflies* (1988) by Isao Takahata (but don't watch it if you have a weak heart). In China, Haber-Bosch plants were in use since the 1930s, but they weren't sufficient to provide fertilizers for the large-scale Chinese agriculture and couldn't avoid the famine of 1960-1961; the last recorded famine in Chinese history.

It was only after the end of WWII that the new fertilizers started having a global impact. The "Green Revolution" was the result of already ongoing trends, but it pushed agriculture to even larger yields. Fertilizers were the equivalent of amphetamines for human beings: they made their targets much more active. One of the main protagonists of the revolution, Norman Borlaug, bred dwarf wheat varieties—short, sturdy stalks that didn't flop under heavy grain heads. The trick wasn't just the lower height: the grains matured fast, letting farmers squeeze two crops into a year. Using this variety, by the 1960s, Mexico switched from a wheat importer to an exporter. The model spread. India and Pakistan, hit by famines in 1961, adopted high-yield rice and wheat varieties. The International Rice Research Institute in the Philippines generated a rice termed IR8, dubbed "miracle rice," tripling yields by 1967. From there, the technology moved to Southeast Asia, parts of Africa, and Latin America. By 1980, global fertilizer production neared 80 million tons annually, mostly nitrogen-based, with the rest phosphate and potash. China joined late, but by 2000, it consumed a third of the world's supply. Fallow periods shrank; two or three crops yearly became the norm where water flowed. Marginal soils—sandy flats, and exhausted tropical soils turned productive.

Fertilizers were not the only factor that created the Green Revolution. Mechanization, using energy powered by fossil fuels, made it possible to till the fields rapidly and without the need for human or animal labor. It also made it possible to transport food all over the world by rail and container ships, preventing local famines. Modern chemistry also created pesticides that made crops resistant to insects and fungal infestation. To say nothing about refrigeration, which allowed perishable food to be stored much longer than before. By the 1970s, the Green Revolution was a global doctrine, and Norman Borlaug received the 1970 Nobel Peace Prize. By 2000, fertilizers underpinned 40–60% of the global food supply, as FAO estimated. Today, about half of the people in the world can eat every day only because of fertilizers.

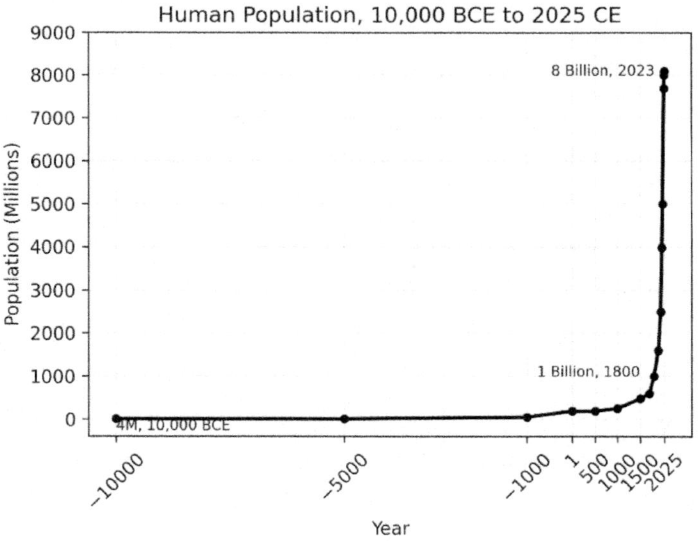

Figure 4 - Human population trends. Data from Our World in Data

The effect on social structure was profound. Surplus food production could feed large cities, creating the phenomenon called "urban sprawl." Asia's population doubled from 1950 to 1990, sustained by the Green Revolution. The massive growth of population in formerly poor countries generated a global power shift. The 19th century was the "British century," and most of the 20th century can be called the "American Century." Now, we are seeing the start of what could be the "Asian Century," propelled forward by the huge Asian populations. Africa is waiting for its turn, with its population also rapidly growing, sustained by mechanization and fertilizers.

The current trends in agriculture involve genetically modified organisms (GMOs). Hailed as the new green revolution, these organisms are engineered to optimize yield and, sometimes, to be resistant to pesticides that would kill natural plants. Then, by the 20th century, factory farms and antibiotics, as well as genetic technologies, turned animal husbandry into a global industry integrated with agriculture, scientifically managed, and optimized for calorie production. These innovations have reduced the need for artificial fertilizers, whose use has been going down worldwide during the past few years, without reducing yields.

The technological success of the new agriculture was spectacular, but not without adverse side effects. Increased productivity came with enhanced fragility. The oil shocks in the 1970s showed the vulnerability of the system. Today, the combination of fossil fuel depletion and climate change promises much larger problems. Soil degradation, water pollution, and nitrates choke rivers and create algal blooms in the sea. The system is vulnerable to the price of fossil fuels: any price increase immediately leads to malnutrition in poor countries. Another effect of the technological revolution is the disappearance of peasants, a social class that had been part of human societies since the beginning of recorded history. Now, peasants have become agricultural workers engaged in operating heavy machinery. They wear hard hats in exchange for their old straw hats. They work fixed schedules, and they receive a monthly salary. When needed, they can be rotated to and from agricultural and industrial jobs.

The latest industrial revolution in food production is "precision fermentation," a form of husbandry in which bacteria are tamed and forced to produce what humans want. Genetically engineered bacteria can produce almost any carbon-based molecule, from special chemicals to foodstuffs that look and taste like the original steak, cheese, butter, or anything humans like to eat. So far, precision fermentation is not a significant source of food for humankind, but that could change in the near future. The first stage of the process will be to use precision fermentation to produce food for livestock, which will then provide food for humans. Later, humans may directly eat food produced by genetically engineered bacteria.

Right now, there is a strong resistance against using precision fermentation to produce food for humans. For instance, in 2023, the Italian government banned the production and import of lab-produced food. The agriculture minister, Francesco Lollobrigida, used the term "Frankenfood" after the farmers' unions collected 500,000+ signatures against lab-grown meat and precision-fermented dairy. The law was mostly symbolic, because there does not exist, so far, a lab-grown food that's approved by the European Union and that can be marketed for human consumption. It remains in force today, but it remains to be seen for how long if a serious food crisis develops, as it may well happen in the near future.

Indeed, despite the success of the Green Revolution and the promises of precision fermentation, these new food production systems are fragile. They largely depend on the availability of cheap oil and natural gas, and if the supply declines, production declines, too. The Green Revolution fed billions, but it was based on finite resources. Precision fermentation is based on high-tech and the availability of energy, too. Neither can continue functioning unless new energy sources are developed and utilized on a large scale.

Norman Borlaug himself noted in his Nobel Prize acceptance speech in 1970 that

> "The green revolution is a change in the right direction, but it has not transformed the world into Utopia. None are more keenly aware of its limitations than those who started it and fought for its success. ... The green revolution has won a temporary success in man's war against hunger and deprivation; it has given man a breathing space. If fully implemented, the revolution can provide sufficient food for sustenance during the next three decades. But the frightening power of human reproduction must also be curbed; otherwise, the success of the green revolution will be ephemeral only."

Borlaug was still looking at the world with Malthusian lenses; he couldn't have imagined that the "frightening power of human reproduction" would have been stopped in its tracks by the decline in fertility. More than five decades after his speech in 1970, agriculture is still able to feed a human population that is no longer growing exponentially as it was doing in the 1960s. It was another manifestation of how unexpected the future tends to be. But will population decline be fast enough to eliminate the risk of new famines? Only time will tell.

The Great Population Inversion

At the end of the last ice age, about 10,000 years ago, the world's population was perhaps one million people. With the development of agriculture about eight thousand years ago, the human population literally exploded. During the past 2,000 years, it has been growing at a dramatic rate. In the 1970s, people started being worried about this apparently unstoppable growth. At that time, calculations were made to show that it was a trend impossible to maintain for long. It was the return of Malthus' ideas and of the concept of the unstoppable "urge to reproduce." Something would stop population growth before it consumed every bit of food on the planet, turning it into a desert. Someone calculated that, extrapolating the observed exponential growth in the 1960s, it would have taken little more than a millennium for humankind to reach the total mass of Planet Earth for a total of 85 sextillion people. Impossible, of course, and some monstrous catastrophe would occur much before it could happen. But that didn't seem to be happening during the 20th century. Instead, something completely different was taking place. It was the slowdown of birth rates called the "Demographic Transition."

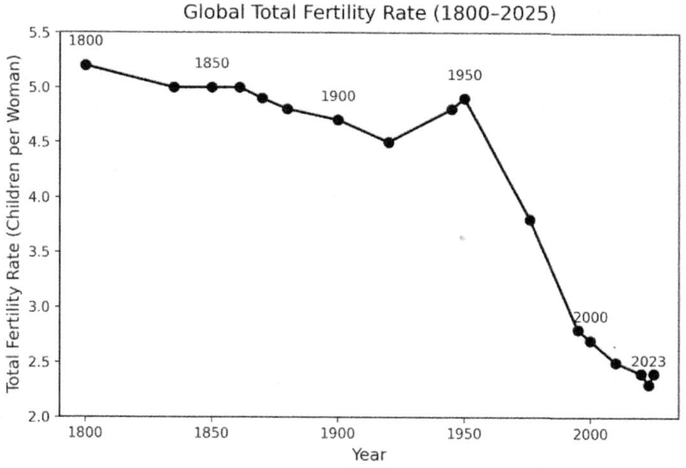

Figure 5 – Global Total Fertility Rate (number of children per woman over lifetime). Data from Our World in Data, UN DESA, and Bairoch 23.

You can see in the figure how the global "Total Fertility Rate" (TFR), the number of children per woman, plummeted starting in 1950. Note that the TFR is not simply the ratio of children to women. It is the result of a calculation that uses demographic data to estimate how many children the current "cohort" of fertile women can be expected to have. An alternative possible metric is the birth rate, the number of children born per one thousand people, but it gives similar results. Something happened in the 1950s that made the global fertility rate plummet to average values only slightly above the minimum level necessary to keep a stable population, that is, 2.1.

If the trend continues, it will soon cross that line, and humankind will not give birth to a number of children sufficient to replace the older generations. And that necessarily means that the population will decline. That's not going to happen tomorrow. It is because the population has a certain "inertia" that makes its trajectory of growth or decline sensitive to previous events. Even though the average fertility rate today is nearly equal to the replacement rate, the global population is still increasing. It is because the old generations are still around, while the cohorts of young women born in previous times now have a chance to have children. In practice, the growth is slowing down, but it has not stopped yet. The trend will reverse only when the number of deaths is larger than the number of births.

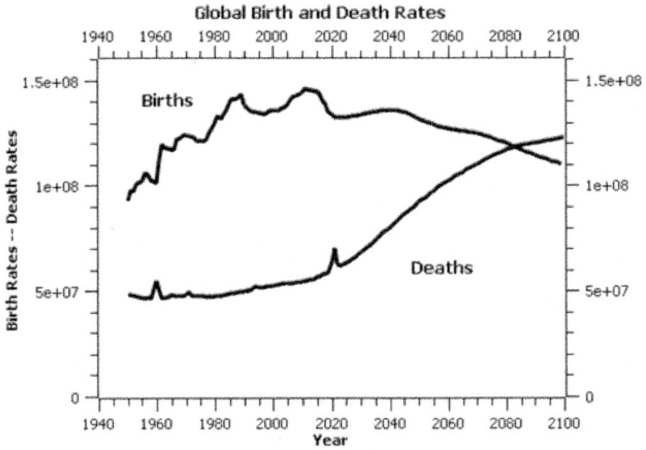

Figure 6 - Global Deaths and Births according to United Nations Data and projections

You see in the figure how, according to the UN data and demographic models, this is not supposed to happen before ca. 2080. The result is that the global population won't stop growing until that date. Today, the global birth rate is approximately 18.5 per 1,000 people, and the global death rate is around 7.8 per 1,000 people. This means that in 2024, there were roughly 146 million births and 57 million deaths annually worldwide. Births outnumbered deaths by more than a factor of two. It follows that, even if, for some reason, the global birth rate were to halve starting today, the world population would still keep increasing. To stop growth immediately, it should collapse to about one-third of the current value. Conversely, if the death rate were to double all of a sudden, the population would continue growing, assuming that only people beyond their active reproduction rate were to be affected. This calculation nicely illustrates the "inertia" of the population trends. A typical result of the UN projections is shown in the figure

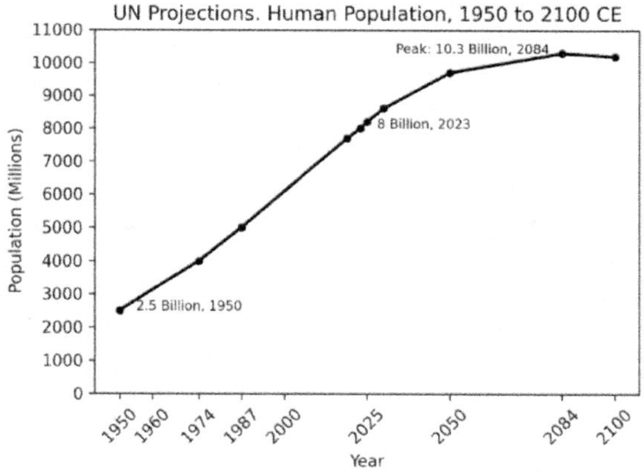

Figure 7 -- One of the UN projections for the global population trends

The demographic models of the United Nations take into account this inertia and project to a relatively far future the stabilization or the start of the decline of the global population. Here is a list of the main results of the UN demographic model calculations during the past decades. [24]

Date	Max Population Year	Peak Population	Population in 2100
2004	2075	9.2	9.1
2012	2100s	No peak	10.9
2015	2100s	No peak	11.2
2017	2100s	No peak	11.2
2019	2100s	No Peak	10.9
2022	2086	10.4	10.4
2024	2084	10.3	10.2

Table 1 – Evolution of the UN estimates for human population

Other demographic studies provided similar results. In the 1950s, Walter Greiling estimated a population of about 9.0 billion by the early 21st century.[25] Lutz et al., at IIASA generated several forecasts that tended to cluster around 9 billion people by 2070. [26,27] Overall, all the studies see the human population peaking in the late 21st century, or even in the early 22nd, followed by a stabilization or a slow decline.

However, as I mentioned in the introduction, these extrapolations are based on statistical models that take into account only the historical trends. Unfortunately, the problem with extrapolations is that the future is never exactly like the past. So, the world population could start its decline much earlier than 2080 if we take into account the various factors that could accelerate the economic decline of the industrialized countries. For instance, chemical pollution and climate change can increase the death rate, while fertility could go down faster than expected. The models based on "system dynamics," that is, based on an integrated balance of the various factors involved in population trend, always indicate a much earlier start of decline. In the introduction, I mentioned the early results of the "Base Case" scenario of the 1972 study The Limits to Growth, which generated the start of the decline approximately in 2050. Later calculations based on similar models confirmed this result and moved the population peak to earlier dates. A recent recalibration of the model generates the population peak a little before 2030.[28] Other dynamic models have added different parameters not purely based on physical constraints, for instance, socioeconomic factors and technological progress.[29] Some of these studies have been published as reports to the Club of Rome.[30], [31] The most recent one, *Earth for All* (2022), [31] generated the population peak around 2030-2040. There are many other qualitative and quantitative studies. Still, most of them place the start of the decline well before the end of the current century, in many cases before 2050.

As is always the case, exact predictions are impossible for complex systems. But it is possible to predict general trends, and it is clear that the end of population growth is going to occur during the 21st century. We can list some of the results of these models in a table.

Year	Study / Scenario	Model	Peak size (bn)	Peak year
1971	Forrester [9]	World 2	5.5	2020
1972	Meadows et al. [10]	World3	7	2050
1992	Meadows et al. [32]	World3	7.5	2030-40
2004	Meadows et al.[33]	World3	7.3-7.8	2030-2035
2012	Randers [30]	Simplified World3	8.1	2040
2021	Bystroff[29]	World 4	7.5-8.5	2025-2040
2022	Dixon et al. [31]	Extended World3	8.5	2040
2022	Dixon et al. "Too Little Too Late" [31]	Extended World3	8.6	2050
2022	Dixon et al. "Rocky Road" [31]	Extended World3	8.4	2042
2024	Nebel et al. [28]	Stock-driven SD	8.1	2025-2027

Table 2. Results for the population parameter by system dynamics models. Forrester. 1971[9], Meadows, 1972, Meadows 1992[33], Meadows, 2004[32]. Bystroff. 2021[34], Randers, J. 2012[30], Randers, 2022[35], Nebel, 2024[28]

Taking a different view, we could examine the planetary trends in light of what's happening in individual countries. There, we find large differences. Almost all industrialized countries have fertility rates well below the replacement level, in some cases close to one child per woman. For almost all of them, the native population is declining, even though the total is often compensated for by immigration. Conversely, many poor countries, and in particular African ones, still have a high fertility rate, with an average of about 5-6 children per woman. But these numbers are starting to decline as these countries are following the same trajectory as the industrialized countries, only with a delay of a few decades. Let's now place all the countries whose populations are declining in a single table. They reported in table 3 (next page) ordered to start with those with the largest population, with China at the top.

These 32 countries represent a total of 2.22 billion people, that is 27.2% of the global population (8.19 billion). In terms of GDP, the 32 declining countries in the table generate $46.815 trillion/year, nearly half of the world's GDP (ca. $107 trillion/year), more precisely 40.7%. In terms of per capita comparison, these countries average $21,042 per person, which is to be compared with the world's average of $14,042. These declining countries are substantially richer than those that show a population increase. In terms of economic power, 40.7% of global GDP reflects a concentration of wealth accompanied by labor shortages, aging people, and more effects on the economy.

The fastest declines occur for some Eastern European countries, Bulgaria, Latvia, Serbia, and Lithuania, with Ukraine having the steepest decline. In all cases, it is caused by a combination of a decline in birth rates and emigration. Southeastern Eurasian countries have higher birth rates (Thailand, South Korea, Taiwan, and China), but they are declining nevertheless. China's population is falling by more than one million people per year. All these countries have total fertility rates below the replacement rate (2.1), with South Korea having the lowest in the list (0.8). Western Europe is experiencing a slower but steady decline, only in part compensated by immigration. These local trends contrast with global growth, driven by Africa and some regions of Asia. We see here a demographic divide that will reshape economic and political landscapes for decades to come.

No.	Country	Pop (M)	Growth Rate	Birth Rate	Death Rate	Total GDP (USD B)
1	China	1408	-0.01	6.77	7.87	18585
2	Japan	146	-0.43	9.8	14.1	1898
3	Russia	124	-0.51	6.6	11.7	4191
4	Germany	84.4	-0.27	9.1	11.8	4456
5	Thailand	71.8	-0.02	8.5	8.7	560
6	Italy	58.9	-0.53	6.8	12.1	2250
7	South Korea	51.7	-0.23	5.1	7.4	1840
8	Spain	47.4	-0.26	7.1	9.7	1569
9	Ukraine	41	-0.87	6.5	15.2	213
10	Poland	37.8	-0.26	9.2	11.8	847
11	Taiwan	23.6	-0.13	6.8	8.1	812
12	Romania	19	-0.50	8.8	13.8	331
13	Cuba	10.94	-0.12	9.9	10.39	108
14	Czech Rep.	10.5	-0.14	9.8	11.2	320
15	Greece	10.4	-0.36	7.6	11.2	241
16	Portugal	10.3	-0.38	7.7	11.5	281
17	Hungary	9.6	-0.38	9.5	13.3	212
18	Belarus	9.2	-0.37	9.3	13	73
19	Bulgaria	6.8	-0.73	8.1	15.4	106
20	Serbia	6.6	-0.58	8.9	14.7	75
21	Finland	5.5	-0.20	8.2	10.2	284
22	Slovakia	5.4	-0.14	9.6	11	133
23	Croatia	3.8	-0.45	8.7	13.2	81
24	Georgia	3.67	-0.13	11.7	13	32
25	Bosnia H.	3.2	-0.28	8.2	11	27
26	Armenia	2.8	-0.03	11.5	11.8	21
27	Lithuania	2.8	-0.45	9	13.5	81
28	Moldova	2.5	-0.38	10	13.8	17
29	North	2.1	-0.08	10.2	11	15
30	Slovenia	2.1	-0.33	8.5	11.8	72
31	Latvia	1.8	-0.53	9.2	14.5	42
32	Estonia	1.3	-0.29	9.8	12.7	40

Table 3. The 32 countries that show population decline in 2024.

Note that the Table shows the Growth rate without distinguishing between changes due to birth/death rates and immigration/emigration. Taking this caveat into account, we see that the fastest declines occur for some Eastern European countries, Bulgaria, Latvia, Serbia, and Lithuania, with Ukraine having the steepest decline (.87% yearly). In all cases, it is caused by a combination of a decline in birth rates and emigration. Southeastern Eurasian countries have higher birth rates (Thailand, South Korea, Taiwan, and China), but they are declining nevertheless. China's population is falling by more than one million people per year. All these countries have total fertility rates below the replacement rate (2.1), with South Korea having the lowest in the list (0.8). Western Europe is experiencing a slower but steady decline, only in part compensated by immigration. These local trends contrast with global growth, driven by Africa and some regions of Asia. We see here a demographic divide that will reshape economic and political landscapes for decades to come.

How about the other side of the world, the growing countries? In Table 4 (next page) you can find a list of the fastest-growing ones. They are listed with the largest countries at the top. To make the table manageable, I included only the African Sub-Saharan countries, which occupy the first 39 places in the list. Other agricultural countries in Asia, Latin America, and other continents tend to have lower fertility rates, although their population is growing.

Rank	Country	Population Growth Rate (%)	Total Population (2025)	GDP per Person (USD)	Total GDP ($B)
1	Nigeria	2.2	230,069,000	2152	497.3
2	Ethiopia	2.6	130,000,000	1294	168.2
3	DRC	3.3	108,059,086	673	72.7
4	Tanzania	2.9	70,559,089	1327	92.9
5	Kenya	2.2	57,009,089	2007	111.4
6	Sudan	2.3	50,090,080	1102	55.1
7	Uganda	2.9	50,069,059	1072	53.6
8	Angola	3.1	38,000,000	2686	102.1
9	Mozambique	2.8	36,000,000	658	23.7
10	Ghana	2.3	35,090,069	2328	81.5
11	Madagascar	2.5	31,579,050	554	17.5
12	Côte d'Ivoire	2.3	31,000,000	2725	84.5
13	Cameroon	2.7	30,000,000	1717	51.5
14	Niger	3.8	28,559,069	615	17.5
15	Burkina Faso	2.5	24,090,090	893	21.4
16	Mali	3.1	24,090,090	879	21.1
17	Malawi	2.7	22,590,059	632	14.2
18	Zambia	2.8	22,000,000	1398	30.8
19	Chad	3.1	20,000,000	740	14.8
20	Senegal	2.6	19,090,090	1896	36
21	Somalia	3.1	19,000,000	592	11.2
22	Zimbabwe	2.1	17,000,000	1697	28.5
23	Guinea	2.6	15,000,000	1189	17.8
24	Benin	2.8	14,090,090	1449	21
25	Rwanda	2.4	14,500,000	1045	15.2
26	Burundi	2.8	14,090,090	238	3.3
27	South Sudan	3.1	13,000,000	400	4.8

Rank	Country	Population Growth Rate (%)	Total Population (2025)	GDP per Person (USD)	Total GDP ($B)
28	Togo	2.3	9,500,000	1013	9.6
29	Sierra Leone	2.5	9,000,000	434	3.9
30	Eritrea	2.3	6,800,000	582	4
31	Rep of Congo	2.4	6,500,000	2214	14.4
32	Centr Afr Rep	2.3	6,000,000	492	3
33	Liberia	2.3	5,800,000	800	4.6
34	Mauritania	2.5	5,000,000	2337	11.7
35	Gambia	2.4	2,800,000	1013	2.8
36	Gabon	2.1	2,500,000	8873	22.2
37	Guinea-Bissau	2.3	2,200,000	1029	2.3
38	Eswatini	2.2	1,800,000	6788	12.2
39	Djibouti	2.1	1,200,000	3873	4.6

Table 4: The 39 African countries that show population growth in 2024. The growth column takes into account the overall population change, which is affected not only by birth and death rates, but also by emigration.

The total population of these 39 countries is approximately 1.17 billion people. These countries account for roughly 15% of the world's population (~8.2 billion). Their combined GDP is $1.93 trillion, which is smaller, for instance, than that of Germany alone. In total, it is less than 7% of the world's current GDP. In terms of GDP per person, we have a yearly total of ca. $1,644 per year, a little more than 10% of the world average. About the countries not listed in the table, most are still weakly growing but are soon going to start their declining trajectory, for instance, India and the US. So, the decline in population will soon be global.

Despite the parallel trajectories, the projections show that there will be a period of a few decades in which the African countries will keep increasing their population, while that of the Northern countries will continue its decline. According to the UN data, we can calculate that by the end of the 21st century, the Sub-Saharan African population could go from the current level of about 15% of the world population to reach about 40%. At the same time, the industrialized world would decline from the current level of 27% down to about 15%.

Projections are always unreliable, but the trend is qualitatively correct. The result has been termed the "African Century," with African countries dominating the world just as European countries did during the 19th and 20th centuries. Not so surprising: after all, a few million years ago, Africa was the only place that hosted a hominin population. It was our cradle, and to that cradle we might return. But note that it is not at all obvious that it will happen, because things change fast in the world, and complex systems always surprise us. However, assuming it is going to happen, at least in some measure, what could the consequences be?

For one thing, the Africans could expand their population beyond their continent, "back-colonizing" Europe and North America. If that were to happen, they could hardly do worse than what Europeans have done in their past trajectory of world domination in terms of exterminations, enslavements, large-scale population displacements, and cultural erasures. Africans moving into the lands of their former colonizers could wisely apply the concept of Ubuntu, often translated as "humanity towards others." But this movement would not necessarily be a form of colonialism. African immigrants could be essential to help the declining Eurasian and North American industrialized states maintain their social and economic structure.

Nevertheless, if history often rhymes, it never repeats itself exactly. The great European colonial expansion of the 19th century was possible in a world that still had abundant mineral resources that could be used to create new military technologies and transport large amounts of people and equipment all over the world. The British

Empire, it is said, was built by wooden ships and iron men. But neither would have been available without the abundant coal resources of England. Other European countries, mainly Germany and France, built their empires in the same way. When the European mineral resources were exhausted, the European armies were used to control the exploitation of the mineral resources of the colonies for the benefit of Europe. A future African World Empire would find little interest in invading the countries of the North, deprived of the mineral resources they once had. So, a primary reason for new wars would no longer exist. In the long run, the global fertility decline would make empire-building impossible. Not that people wouldn't be able to find reasons to kill each other, but at some point, the age of colonial empires will be over forever. One thing is certain: The great population inversion is coming!

4. The Limits to Population

"The perpetual tendency of the race of man to increase beyond the means of subsistence is one of the general laws of animated nature, which we can have no reason to expect to change." Thomas Malthus, "An Essay on the Principle of Population" (1794)

Figure 8 - The Four Horsemen of the Apocalypse. Victor Mikhailovich Vasnetsov, 1887.

In the previous chapter, we saw the factors that allowed the human population to grow to the present levels – more than eight billion people. In this chapter, we examine the factors that can limit population growth and that, in principle, can be expected to limit the number of people who can live on Earth in the future. We can take inspiration from the mention of the four horsemen of the apocalypse mentioned in the Bible: Death, Famine, War, and Conquest. Here, however, I interpret them differently, using the names of some of those who discussed the subject. The knights are Thomas Malthus (famine, war, and plague), Edward O. Wilson (socioeconomic factors), John Calhoun (stress), and Rachel Carson (pollution). I don't mean, of course, that these people were advocating death and destruction for humankind! Their names are just convenient labels for different interpretations of the same phenomenon. It should be taken into account, in any case, that we are dealing with a complex system, that of the trajectory of the human population on Earth, with all the factors, limits, obstacles, side effects, and feedback involved. There is no chance that a single explanation can account for everything that happened and will happen. But from reviewing the factors that place a limit on the number of human beings, we can gain some insight into what to expect.

Malthus: Famine, War, and Plague

. Few people in history have been so badly misunderstood as the Reverend Thomas Malthus (1766 – 1834). Just as an example, if you have a chance to watch Werner Boote's documentary *Population Boom* (2014), you may notice that it starts by quoting Malthus, who is said to have stated that "a huge famine will strike the world in 1860." Since no such famine occurred, the narrative of the movie suggests that Malthus was wrong in everything he said. A little questionable as a syllogism, especially considering that Malthus never said any such thing. Werner Boote is a wonderful person whom I had a chance to meet in person. His movie is also wonderful despite the initial blunder. Too bad that he fell into the common mistake of believing in the pervasive myth that sees Malthus as not just a doomer, but a doomer of the worst kind, being both wrong and evil. But, if that's what he was, why are we still discussing his work, even if only to pelt him with metaphorical rotten tomatoes?

In reality, Malthus was a pioneering thinker whose ideas have influenced several scientific fields, including economics and evolutionary biology. To be fair, he was not the first to note that there are physical limits to humankind's growth. Before him, Benjamin Franklin and Reverend Otto Diederich Lütken, a Danish cleric, published texts in which they noted the problem. Yet, it was Malthus who remained in people's memory as the originator of population studies with his book An Essay on the Principle of Population, published in 1798 and then in a revised version in 1803. Darwin himself acknowledged his debt to Malthus, and a whole generation of classical economists, from David Ricardo to Adam Smith, took from him the concept of "scarcity" as the basis of their theories. An influence that remains visible to this day.

The originality of Malthus' ideas was not just to note that negative checks on population growth exist, but also that the problem could be quantified in mathematical terms. The idea that the universe follows mathematical laws was first developed by Galileo Galilei during the 17th century, but, up to Malthus's time, had been applied only to "hard" sciences; that is, physics or engineering. Malthus was no mathematician; he couldn't even remotely match Newton's skills with equations and calculus. Still, he was among the first to apply mathematical laws to a "soft" subject, as we call those fields of science that deal with social factors. Before Malthus, mathematics was only applied to "hard" fields of science, that is, physics, chemistry, engineering, and the like. His key point was that population grows geometrically (1,2,4,8,16...), while food production can only grow arithmetically (1,2,3,4,5...). The obvious consequence was that population growth would always outstrip food production. And the result could only be famines and other disasters.

A good summary of Malthus' ideas can be found in this paragraph from chapter 7 of his 1798 book.

> "The power of population is so superior to the power in the Earth to produce subsistence for man, that premature death must in some shape or other visit the human race. The vices of mankind

are active and able ministers of depopulation. They are the precursors in the great army of destruction; and often finish the dreadful work themselves. But should they fail in this war of extermination, sickly seasons, epidemics, pestilence, and plague, advance in terrific array, and sweep off their thousands and ten thousands. Should success be still incomplete, gigantic inevitable famine stalks in the rear, and with one mighty blow levels the population with the food of the world."

It is more than a little overdramatic, but Malthus makes his views clear. Note how he attributes the problem to the "vices of mankind," which we can interpret as the sin of fornication, a popular concept at Malthus' time. The note on how these "vices" can "finish the dreadful work" refers to venereal diseases, particularly syphilis and gonorrhea, significant public health concerns at the time. But the real population controllers were epidemics, plagues, pestilence, and, eventually, famine. Malthus was simply reporting what he could observe in the world around him. In another paragraph from the 1803 edition, he also mentions war as a factor that checks population growth.

> "War, that great engine of destruction, is a check to population, which operates with more or less force in almost every country. It sweeps off many thousands in the prime of life, and by the miseries it occasions, prevents many more from coming into existence."

Note, among other things, how Malthus states that the catastrophes would "level the population with the food of the world." That is, he says that famines *stabilize* the population, and he never mentions the possibility of a population collapse. You can examine Malthus' works, but you will never find anything that you can understand as meaning that "on such a date, a catastrophic famine, or plague, or whatever, will occur." So, Malthus' ideas were nothing comparable to those that our modern "collapsniks" peddle all the time. In other words, Malthus was not a catastrophist!

But curiously, this is the main accusation against him, that is, of having made catastrophic predictions that turned out to be wrong. The human mind not only tends to forget things but, much worse, to remember them wrong. Today, we would consider Malthus not just a non-catastrophist but actually an optimist. In his work, he never mentions limits to the extent of the area that could be utilized for food production. It must have been obvious to him that these limits had to exist, but they appeared so remote that he thought they could be ignored. It was a view similar to that of building developers today, who usually don't discuss, and perhaps don't even conceive, the ultimate limits to the extent of land they can cover with concrete. Of course, Malthus couldn't imagine that technological improvements would increase the productivity of agriculture to the levels of today, when it supports more than eight billion people. But he had correctly framed the problem. Better technology can expand the limits to population growth, but the limits remain.

But why exactly do humans always tend to push against the physical limits of what their environment can do? Malthus, just as many others of his time, spoke in terms of the "urge to reproduce," the human tendency to have as many children as possible. For Malthus, a Christian cleric, this urge was strictly related to the unchecked desire for sex, and hence famines could be seen as God's punishment for people's bad behavior. There is a certain degree of contradiction between God's command to his people to "grow and multiply" and chastising the activity that makes that possible. But this is not a book on theology, and so let's skip this subject.

By looking at famines and other disasters in mathematical terms, Malthus was on the right track to understanding the role of resources in the economy. But he, just like the early economists, missed a fundamental element of the behavior of real-world markets:

the factor we call today "overexploitation." It is a concept that became clear only during the 20th century, and it was initially understood to occur only with exhaustible resources, such as mineral resources. Only in later times would it be understood that it affected agriculture and food production as well. Even traditional agriculture, commonly seen as a gentle, ecological activity, is subject to depletion because it depletes the soil of nutrients. Modern industrial agriculture, then, is even more sensitive to nutrient depletion because it relies on fossil fuels, an obviously finite resource.

The British economist William Stanley Jevons (1835 – 1882) arrived on the scene a few decades after Malthus, and he was the first to propose a quantitative model of resource depletion. His book The Coal Question (1865) was a milestone in the field, an analysis that is still perfectly valid today. Jevons' basic idea was that a high-grade resource is concentrated and hence easy to extract at a low cost. Obviously, the least expensive resources, the high-grade ones, are those extracted first. Then, the industry must gradually move to more expensive, lower-grade resources. Jevons noted that, eventually, the cost of extraction would force producers to raise prices to levels that consumers couldn't afford. This would reduce demand and, hence, production. Eventually, the demand would be reduced to zero, and extraction would stop, even before the resource was completely extracted. In other words, the limits to a resource are not given by its geological amount but by its cost of production. A fundamental intuition which, unfortunately, people tend to forget, even today. Replace "coal" with "fertile soil," and you see how Jevons's ideas are perfectly applicable to agriculture.

Later, in 1931, Harold Hotelling arrived at similar conclusions with the model we call today "Hotelling's rule."[34] However, he added to his model a factor that Jevons hadn't considered: substitution. Hotelling's idea was that before a finite resource ran out, its increasing prices would make it profitable to use an alternative resource, the "backstop" one, so that the supply to the market would be assured, although at higher prices. This element made Hotelling's model look more optimistic than Jevons' one, and that was the probable reason why it is considered the standard treatment of depletion in economics textbooks. But Hotelling's model is not optimistic at all. Instead, it looks like Marie Antoinette's statement that if the poor run out of bread, then they can eat cake, with the implication that if they can't afford to pay for it, well, it is their fault. Incidentally, just like Malthus, Marie Antoinette was reviled for something she never said.

Modern economists, however, often seem to reason like Marie Antoinette never did. They use a sleight of hand that has a name: the "Functional Model."[35] It assumes that increased prices generated by depletion will stimulate technological progress, and that will, in turn, lower the prices of the backstop resource to the level of those of the exhausted one. Then, production can continue, and everything will be well in the best of worlds. It was as if Marie Antoinette had reassured the people who were taking her to the gallows that better technologies would soon enable Parisian bakers to sell cake at the same price as bread.

These optimistic views were challenged in recent times by a group of economists, sometimes defined as "bioeconomists,"[36,37] who believe that depletion cannot be swept away on the basis of the idea that technology can always find a substitute for every resource depleted beyond profitability. But much before the concept of bioeconomy became popular, a new theory of depletion was put forward not by an economist but by a geologist: Marion King Hubbert. It is typical that when a scientific field becomes stuck in an untenable set of ideas, the only way to push it forward comes from people working in different fields who sweep away the old ideas, introducing new ones. Perhaps Hubbert didn't know what a backstop resource is, or maybe he didn't care. In any case, in 1956, he proposed the idea of a bell-shaped production curve for crude oil. [38]

Hubbert's model was completely independent of economic concepts such as prices and markets. It was based only on empirical observations, and it posited that production initially grows at increasingly fast rates, then it starts slowing down, reaches a peak, and finally starts an irreversible decline. The "Hubbert Curve" turned out to be a general model applicable to almost all minerals produced by the mining industry. It was found to be applicable also to agricultural and other food resources, such as fisheries, which tended to overexploit and destroy the resources they used.

Hubbert's model turned out to be steeped in population biology, and it could be reproduced using the same mathematical models. [39] Its application to renewable resources was analyzed, perhaps for the first time in 1968, by Garrett Hardin, a biologist, in a paper titled *The Tragedy of the Commons*.[40] The "commons" is an ancient tradition existing everywhere in the world: it is a term that defines a resource that's free for all to exploit. It can be pasture, wood, mushrooms, fish, and more. Hardin described his model in terms of the pastures in England in the 18th century. He started from the observation that in the economy of common pastures, every shepherd is free to bring to graze as many sheep as he wants. And that is where the problem lies. Imagine you are a shepherd, and you are also a good economic agent, in the sense that economists use the term. Of course, you want to increase the yield of your activity, and the way to do it is to have as many sheep as possible. Hence, your flock becomes larger and larger as time goes by. At the same time, every shepherd in the area is doing the same thing: enlarging their flocks.

The problem arises when the sum of all the sheep reaches the limits of the pasture to nourish them. The result is nothing like the stabilization that Malthus had proposed in terms of "leveling of the population with the food." For a while, the total number of sheep may continue to increase, but they are overgrazing the pasture, leaving new plants no time to grow before they are eaten. As a professional shepherd, you know perfectly well that adding one more sheep to your flock damages the pasture. Still, if you are also a profit-minded agent, you know that the extra sheep will significantly increase your profit. So, rationally speaking, you should add that sheep to your flock. And here lies the tragedy. Every shepherd will think along the same line, and soon, the number of sheep will grow so much that the pastures will become brown and turn into a desert, perhaps forever if the disaster happens on a large scale. Later, Hardin applied his model to human overpopulation in his book *The Ostrich Factor: Our Population Myopia* (1999).

Hardin's work was a simplification of a more complex reality. Later, Elinor Ostrom, among other things the first woman awarded the Nobel Prize for economics, showed that in the real world, social mechanisms exist in pastoralist societies that prevent the destruction of the commons when they are seen as something to be shared by a community.[41] Nevertheless, Hardin's mechanism can be observed in other kinds of commons, such as with large-scale fishing. Ilaria Perissi and I discuss overexploitation in fisheries in our book *The Empty Sea* (2020).

So, it doesn't matter whether a resource is renewable or not; overexploitation is a universal phenomenon that happens whenever the re-forming rate of the resource being consumed is significantly lower than the extraction rate. Mineral resources are simply a case where the "reproduction rate" involves millions of years. In the case of pastures, grass takes weeks to regrow, but it is eaten by sheep in hours. The same is true for fish, which reproduce in times of the order of years, but they may be removed from the sea in days. The final result is the same: the resource disappears, and the economic activity that exploits it collapses.

You could argue that agriculture is different than pastures. After all, farmland is not managed as a common resource; it is someone's property. Hence, farmers are not supposed to destroy their fields by overproducing. As a result, many economists saw privatization as the solution to the problem of overexploitation. But things turned out not to be so simple, and people are known to overexploit their possessions when they privilege short-term gains over long-term ones. This was demonstrated by field studies by the Norwegian economist Erling Moxnes.[42] In any case, even if you assume that privatization is the solution, it is not always practical, or even possible. Pastures occupy large swaths of land, and they are not easily fenced. Besides, how do you fence the ocean? There exist elaborate schemes that attempt to create "virtual fences" to stop overexploitation in fisheries, but they are known not to work very well.

The Tragedy of the Commons described by Hardin is an abstract model, but it well describes the problem of depletion in agriculture. The difference with pastures is that the soil degradation by agriculture is a slow process that doesn't lead to an immediate disaster, and farmers may miss what's happening to their fields. Soil degradation was studied for the first time in scientific terms by none other than Charles Darwin. He found that soil reforms itself by natural processes, but very gradually and in times of the order of decades and centuries. Soil is the real "skin" of Earth, a machine that continuously recycles nutrients and keeps the above-ground plants alive. But to maintain its capability of supporting agriculture, it must be continuously supplied with dead plant and animal matter that its fungal fauna recycles into nutrients that are used by new plants.

Humans, instead, tend to remove plants from the land and use them as food. In part, the minerals contained in food are returned to the land in the form of organic fertilizers. But a lot of it does not return, and the soil gets gradually depleted of nutrients. In addition, when the soil is left unprotected due to a lack of a surface layer of plants, it is subjected to erosion. It can be washed away by rain or wind and lost in rivers and eventually into the sea. When overexploited in this way, soil behaves like a mineral resource: it doesn't last forever.

So, it is no wonder that if farmers keep farming, the soil gradually becomes thinner and thinner, until the peasant will be toiling on little more than a rocky layer. You can find visual evidence of the process in many places in the world. One is the heel of Italy, the region called Apulia. Some parts of the region are renowned for their *trulli*, characteristic stone dwellings with conical roofs. A picturesque landscape, which, however, is in part the result of soil erosion. The local farmers use the stones they removed from their eroded fields to make walls and roofs. Near the trulli, you can also find *specchie*, conical piles of stones accumulated to remove them from the fields.

Similar problems occur for water. A natural ecosystem exists because it has adapted to the local water resources. When a human agricultural system replaces it, rain may not be sufficient, or the damaged soil may not be able to keep enough water available for plants. In particular, deforestation to make space for human crops destroys the capability of the ecosystem to pull water from the sea in the form of water vapor using the mechanism called the "biotic pump." This pump is a fundamental mechanism that keeps the inland areas of continents well supplied with water, transported there by the "atmospheric river," whose movement is controlled by forests. This mechanism was discovered in 2006 by two Russian scientists: Viktor Gorshkov and Anastassia Makarieva.[43]

To supply enough water to the fields, humans have to rely on artificial irrigation, pumping water from rivers, lakes, or wells. But that may create more erosion and carry soil away faster. In addition, in many cases, water is extracted from wells that are connected to underground aquifers, geological formations capable of storing groundwater. Aquifers are not subterranean lakes but porous rocks where water is stored in tiny pores connected in such a way that it can flow to the surface when it finds a way to move there, as happens when humans dig wells.

Modern aquifers may have formed in remote times and may not be resupplied anymore by rain in amounts sufficient to replace the water extracted for irrigation. In this case, we speak of fossil water generated by ancient rains. The oases in the Sahara Desert are remnants of ancient times when the Sahara was green and experienced significant rainfall. These oases are connected to ancient aquifers, and you can find spectacular waterfalls in the middle of the desert.

Figure 9 -- A waterfall in the Sahara Desert, some 100 km South of Cairo, in Egypt. Photo by the Author,

In practice, agriculture is not an isolated system that can be described by simple models such as Malthus' or Hardin's. It is embedded in the economic system and needs resources that can only be produced by the industry: fertilizers, pesticides, refrigeration, transportation, etc. Agriculture is also directly linked to the human population, and people cannot survive without it. Finally, agriculture is affected by pollution in various ways, including climate change, which affects weather patterns.

Taking all these factors into account is a task that calls for sophisticated models. In practice, what we can learn from both mathematical models and historical data is that agriculture is not forever. It tends to collapse abruptly when the overexploitation limits are overcome. It is a typical example of a Seneca Collapse, the phenomenon that sees slow growth and rapid decline. It is the fundamental cause of the long series of famines that plagued ancient societies over the past several thousand years. Will we ever learn how to avoid it? On this question hinges the destiny of humankind. One thing is clear, though. Malthus was right when he understood that food is a major limit to human expansion. We may have found tricks to circumvent that limit, but that cannot last forever.

Malthus' ideas about population limits were not limited to famines. He mentioned several times at least two more limiting factors: plagues and wars. Both are well known to cause population drops, in some cases, dramatic ones. Malthus knew about the "Black Death," the outbreak of bubonic plague that killed perhaps 30% of the European population during the 14th century. He also knew about the Thirty Years' War in Europe (1618-1648), which, although one of the most destructive conflicts in history, didn't kill more than about 2% of the European population. Malthus didn't see, and probably couldn't even imagine, the genocides of later centuries, such as that of the Native Americans, which killed off nearly 100% of the population of a whole continent. He was, however, correct in seeing famines as the trigger for both plagues and wars.

Today, the question of limits to growth has been examined and discussed from the time of the first quantitative model studies analyzing the possible trajectory of the world's economic systems, for instance, the work by Jay Forrester [9] and that of the MIT group that created the first version of *The Limits to Growth* report in 1972. On the whole, these studies support the idea that non-renewable or slowly renewable resources are a fundamental stumbling block to the increase of the human population. But the modern studies go well beyond the oversimplified Malthus' ideas. Resources are just one of the many elements that interact with each other to force the human population within the limits imposed by the natural ecosystem. How these limits will actually influence the global trajectory of the human population remains to be seen, but there is no doubt that these limits exist.

Wilson: Sociobiology and the Economy of Child Bearing

Edward O. Wilson (1929–2021) is best known nowadays for his sociobiological interpretation of fertility, expressed mainly in his book *Sociobiology: The New Synthesis* (1975), which generated one of the greatest scientific controversies in biology of the 20th century. It was part of the wider debate about evolutionary psychology and the modern synthesis of evolutionary biology. According to Wilson, living creatures behave the way they do in order to maximize the propagation of their genes. This idea was proposed in parallel by Richard Dawkins by using the concept of "selfish genes," which was the title of his 1976 book *The Selfish Gene*. In both cases, fertility is determined by evolutionary constraints.

In sociobiological terms, parents are striving to make their children successful in reproducing themselves and carrying on the parents' genetic heritage. They will try to generate as many children as necessary to ensure that they will survive and reproduce themselves. Critics say that the sociobiological interpretation oversimplifies things; culture, policy, and individual choice can't be reduced to biology alone. The story is indeed complicated, and you can find a comprehensive summary in the recent book by Hopcroft and others, *Not so Weird After All* (2024).

Wilson's sociobiological lens is very much the same as that of Gary Becker (1930–2014). Becker is not very well known today, but he was a pioneer in many fields of economics and was awarded the Nobel Prize in 1992. Among several other contributions, we owe to him the concept of "human capital." One of Becker's key works was his 1960 book, *An Economic Analysis of Fertility*, where he modeled reproductive decisions as rational choices. The idea was that parents weigh the costs (time, money) and benefits (emotional, economic) of having children. To explore this field, he had to move beyond some of the main assumptions of classical economics, that is, that everything that people do, they do in order to make money. A more general concept is that of "utility." It has to do with returns that are not monetary profit, but that move people nevertheless: pleasure, status, well-being, and the like. It had already been explored by the classical economists of the 19th century, but Becker applied it to family planning. In this interpretation, parents make the best choices to maximize the economic success of their children. They try to maximize the capability of the family to invest in their raising and their education in such a way as to optimize their success in life. Just swap "fitness" for "utility," and you arrive at the same result as in the sociobiological interpretation: people tend to optimize their family size in terms of ensuring success for their offspring.

This said, there are several possible strategies to win the reproduction game, both in economic and biological terms. Economists normally tend to strive for a single solution to the problem or optimal resource allocation. Biologists, instead, tend to accept that different creatures use different survival strategies. In terms of reproduction, it is common in biology to speak in terms of two main reproductive strategies: the *r* and the *K*. The terms *r* (lowercase) and *K* (uppercase) are derived from the traditional way of writing the Verhulst logistic equation (see the appendix for details). These terms stand, respectively, for reproductive rate and carrying capacity. The *r* strategy involves having a large number of offspring quickly, with minimal parental investment. In our military-oriented age, we could call it the "carpet bombing" strategy: dropping many bombs and hoping that at least some will hit the target. The *K* strategy, instead, is based on taking care of offspring by means of substantial parental investment. In military terms, we could call it the "sniper's strategy," trying to obtain the maximum result with the minimum effort.

Whether one or the other strategy is the most effective one depends on many factors and circumstances.

Of course, the *r* vs. *K* view is oversimplified, and today it is normally considered obsolete, being replaced by a concept called "Life History." However, it is still helpful as a first approximation to understand the biology of reproduction. There are many examples of species preferring one strategy or the other. Most small animals, such as fish and many insects, play the *r* game, spreading their offspring around. It is a wasteful strategy since only a few will survive, but this waste has to be balanced with the fact that it doesn't require the effort of nurturing from parents. In any case, the life of very small animals is too short and uncertain for the *K* strategy to be worth trying. But, since every rule in biology has exceptions, you may be interested to know that a jumping spider called Nunella (*Nuctenea umbratica*) measures less than 2 mm in body length and yet takes care of its offspring, not unlike much larger animals. [44]

The *K* strategy, instead, is typical of mammals and birds, and it requires a consistent investment of time and resources from the parents in terms of feeding, caring for, and in many cases instructing the young on the ways of life of the species. Of course, animals do not consciously choose whether or not to reproduce, but their strategies are dictated by their genetic setup, shaped and optimized by evolutionary forces. Humans are typical *K* strategists, too. The investment that human parents make in every child is large, possibly larger than in any other species.

Humans are sociobiological K-strategists in the sense that a combination of genetic, cultural, and economic factors dictates their choices in terms of childbirth. Sometimes, it is difficult to understand what exactly is cultural and what is biological in human choices. Still, there is no doubt that humans strategize their reproductive choices according to their perception of the world around them. You just have to take a look at the population curves of a country to note how, after a war, the birthrate usually increases. It is the case of the "baby boom" of the 1950s and 1960s, which affected most of the world. Evidently, families and individuals perceived the need to replenish the losses caused by the war, and also understood that their children had better chances to survive and prosper in a world that had lost much of its previous human capital. So, it is not surprising that humans use flexible and adaptable strategies depending on circumstances: food availability, population density, social status, and others.

Once we arrive at this interpretation, the consequences are straightforward, no matter whether you take the economic or biological viewpoint. For the poor in ancient agricultural societies, children were not expensive and were soon able to help their parents with the farm work. A further bonus was that they would support their parents when they got older. In a way, peasants behave like the shepherds of Hardin's concept of the "tragedy of the commons." Just like shepherds tend to bring as many sheep as possible to the pastures, peasants tend to bring as many children as possible to the farmland. The result was the "unstoppable urge to reproduce" that Malthus and others had identified. It was the r-strategy adopted by human beings.

Modern poor families, even though not peasants anymore, might still reason along similar lines. They know that the investment they could reasonably dedicate to the education of their children is insufficient to give them a serious chance of climbing the social ladder by, say, attending an Ivy League college. So, they might use, at least in part, the *r* strategy and have relatively large families. This behavior is typical of the inhabitants of degraded suburbs surrounding urban centers, often encouraged by state institutions or religious groups that may provide food and money for the children of poor families. Instead, middle-class families can afford to send at least one or two of their kids to college, but no more than that. So, they tend to adjust their family size accordingly, adopting the *K* strategy.

We see this phenomenon most clearly in industrialized Asian countries, such as China, Japan, South Korea, and others. In these societies, high school and college grades lead to the best chances of success in a competitive working environment. As a result, there is a strong competition for the best and most expensive schools. Having fewer children allows families to afford cram schools, which will prepare their child for the competition ahead by repeated drills and exercises. In turn, this method of selection reproduces itself, creating conditions in which a dearth of young people will focus more and more resources on making them the target of more cramming to turn them into highly efficient managers.

Conversely, in developed Western countries, rich families can afford child care and instruction for more than 1-2 children, so they tend to have more children than middle-class families. Sources like the Institute for Family Studies report data on the total fertility rates (TFR, or average children per woman over her lifetime) for the United States:

- Low Income (under $30,000/y): TFR = 2.1-2.3
- Middle Class ($50,000–$100,000): TFR= 1.7-1.8
- Upper Middle Class ($100,000–$300,000): TFR= 1.8–1.9
- High Class (>$300,000): TFR = 1.9-2.0

The very rich, billionaires, are reported to have sired many more children, typically from several women. Elon Musk, for instance, is reported to have had 14 children from four different women, at least up to 2025.

The question of the optimal number of children is inextricably linked to the family structure. In most of the world, the basic idea is that of the "nuclear family," composed of a couple and their children. But just as there are many strategies to optimize the family size, there are many ways to understand how a family can be created. This is a domain typical of biology, where economists do not usually venture. Studies in the field called "Ethology" examine the behavior of non-human creatures in nature, and in particular, how animals organize their reproductive behavior. Many mammal species, and many primate ones, do not form pair-bonded male/female families, but groups where a single alpha male has a large harem of females ("polygyny"). The opposite, single females having several male partners ("polyandry"), is much rarer both in nature and among humans. The two terms, polygyny and polyandry, are described as "polygamy," meaning several sexual partners in formal or informal marriage arrangements.

Some human societies allow certain forms of polygamy, typically polygyny, but most tend to pair-bonding, at least officially. In Western societies, polygamy is forbidden by law. Still, well-off males may practice a form of time-delayed polygyny, also called serial monogamy. They divorce their wife to marry a younger woman, and then repeat the process when the new wife gets older, too. That allows them to have more children than monogamous males – it is a strategy that pays in sociobiological terms, but only for those men who can afford it.

Males are not the only sex that engages in multi-partner intercourse. A persistent legend that haunts the dark corners of the internet is that women are genetically wired for "hypergamy," the social practice of marrying or entering a relationship with one or several partners of a higher social class or status. It involves scanning for upgrades in the male market, prioritizing physical attractiveness (height, jawline, fitness) above all else. Males with these characteristics are called "chads," a term equivalent to that of "alpha males" in biological studies of non-human creatures. The female equivalent of a Chad is called a "Stacy." According to the legend, a woman typically settles for a decent-but-unremarkable guy (the "betabuck," stable job, nice personality, but no spark). Meanwhile, she quietly pines for a Chad. Online folklore insists women are hard-wired for ruthless hypergamy, forever trading up to a taller, richer Chad while keeping a reliable 'betabucks' provider in reserve. This idea has produced its own modern martyr class: that of the "incel," (involuntary celibate), a male who can't find a woman to pair with. Like most internet parables, it contains a grain of truth wrapped in layers of caricature and resentment. Reality is much more sophisticated and nuanced, although the story does have some ground in biological studies.

These trends have to do with the intricate mating game that human beings play all the time. However, for what we are interested in here, demographic trends, male philandering and female hypergamy are just details. What counts is how families, temporary or long-term, gear up to plan for the number of their offspring. Most people love children, and they are happy to have them. But they also want the best possible future for them, and that, clearly, is determined in large part by practical considerations.

These considerations explain, at least in part, the fertility collapse that we have been seeing during the past few decades in industrialized countries. Facing an increasingly competitive social structure and, at the same time, an economic stasis or contraction, families are betting their limited resources on a small number of heirs in order to maximize their chances of being successful. It is the basis of the economic treatment developed by Becker, which goes hand-in-hand with the sociobiological explanation developed by Wilson and other biologists.

Still, we should always be aware that single causes almost never exist in complex systems. Is the economic/sociobiological explanation sufficient, alone, to explain the decline in fertility in the West? There is a nagging problem here. The optimal family size may well be of just one child, but how can it be that many families choose to have zero children? That's surely not the way to optimize one's genetic and economic inheritance.

According to the available US data, about 44% of non-parents aged 18-49 in 2021 said they were unlikely ever to have children, up from 37% in 2018. Among women, 46.5% aged 15-44 were childless in 2022, with higher rates among younger women (97.4% for ages 15-19) and those without a high school degree (68.4%). For men and women aged 40-49, 85% of non-parents reported it was unlikely they'd have children, suggesting a significant portion may remain childless lifelong. Similar numbers are reported for most industrialized countries. Most of these people say that they chose to have no children simply because they didn't want to. The result is couples sometimes defined as "dinks," that is, "double income and no kids."

In terms of economic analysis, "zero children" families can also be seen as a rational choice. In this case, parents judge that their chances to provide sufficient resources for the raising and the education of their children are simply too low to be worth trying. So, they make a different choice. Returning to the sociobiological interpretation, you could say that a nephew or a niece (a "nibling") carries about one quarter of your genes, so that two niblings are the equivalent of one son or daughter. Being an uncle/aunt is much less expensive and troubling than being a parent, and it offers the possibility of passing at least some of one's genes or one's inheritance along.

That might be a correct interpretation, but it is also true that in a society increasingly based on one-child families, even niblings tend to become rare. In practice, many people seem to concentrate the resources they have on themselves and on their individual careers. It is an attitude sometimes named "workism" (not the same thing as wokeism!), another way to optimize one's utility function. It dominates the minds of many people in the West and in East Asia, convincing them that the more one works, the happier one should be. In some cases, it gets to the extreme of what in Japanese is called *karoshi,* death from overwork. It refers to sudden death due to cardiovascular and cerebrovascular diseases like heart attacks and strokes, which are often caused by extreme stress and long working hours. The concept can also include suicide resulting from work-related mental disorders, a phenomenon known as *karōjisatsu*.

Even without resorting to killing oneself by overwork, in many East Asian countries, such as Japan and South Korea, fertility rates have plummeted partly because people (especially women) feel they can't juggle careers and family at the same time. In addition, the fact that women tend to attend school means that they'll delay having children. Sometimes, environmental concerns persuade people that they have to reduce their impact on the ecosystem by having few children or none at all.

These factors are all possible explanations for the plummeting fertility rate, but there is also another interpretation. It may be that people practice the "fox and grapes" strategy. They would like to have children, but they find that they can't, and later rationalize that they didn't really want to have them. Pollution factors, rather than individual choices, may be a major force in the fertility decline, but we will see that in detail in another chapter.

Calhoun: Stress and Overcrowding

John B. Calhoun (1917 – 1995) is still well known today for the studies he conducted on mouse populations in the 1960s and 1970s. While Malthus had proposed that population growth would be stopped mainly by a lack of food, Calhoun explored the opposite case: how population would react to stress generated by crowding, without being constrained by the availability of food.

Obviously, Calhoun couldn't perform his experiments on human beings, but he thought that mice could be a suitable substitute. His setup consisted of an enclosed space, the "mouse utopia," providing unlimited food, water, and nesting material. The version called "Universe 25" began with eight mice (four breeding pairs) in an enclosure of about 6 square meters. The mice started reproducing at a nearly geometric rate until they reached a peak of 2200 individuals after about one and a half years. At that point, they stopped reproducing or generated only sterile offspring. Even before that point, the social structures of the colony had begun to break down. Aggression became common, many females stopped reproducing, and males lost interest in mating. The experiment was stopped when only old and non-reproducing mice remained, prefiguring the impending extinction of the colony. Calhoun explained his results in terms of the stress that affected mice in the chaotic conditions of the overcrowded environment. He termed this phenomenon the "behavioral sink," suggesting that the same factors could affect human society if the world's population were to continue growing, as it was doing in the 1960s and 1970s. [45]

Calhoun's cultural influence was large, and his work is still widely remembered and cited today.[46] It seemed to many that the behavioral sink prefigured an unpleasant human future in an overcrowded world of poverty and stress. It was popularized, fictionalized, and even turned into movies, such as *The Secret of Nimh* (1982). It was a great movie, but its relevance to the behavior of real rats (or men) was not higher than that of Disney's Mickey Mouse movies.

Calhoun's work is remembered today as if it had been unique in the field of population biology. In reality, though, he was just one of the several researchers who studied the behavior of animals when abundant food, but limited space, were available. Among the people working in this field, there were John J. Christian, Charles H. Southwick, and Karl T. Delong. Even earlier, in the 1930s, H. Selye studied the adaptation elements of populations confined in limited spaces. These studies eventually sank into the depths of old scientific literature, even though they are still occasionally cited by specialists in the field.

Remarkably, Calhoun never cited these parallel or earlier studies in his papers in the 1960s and 1970s; a perplexing omission, to say the least. Additionally, the earlier studies were much more detailed and quantitative than Calhoun's studies, mostly based only on qualitative observations. Calhoun tended to interpret the behavior of his mice on the basis of his personal views. For instance, he identified a group of mice that he called the "beautiful ones," characterized by their withdrawal from social interaction, lack of interest in mating, fighting, or other social activities. But there are no published data or publicly available images about this specific group, nor do we know what parameters were considered necessary to classify a mouse as a member.

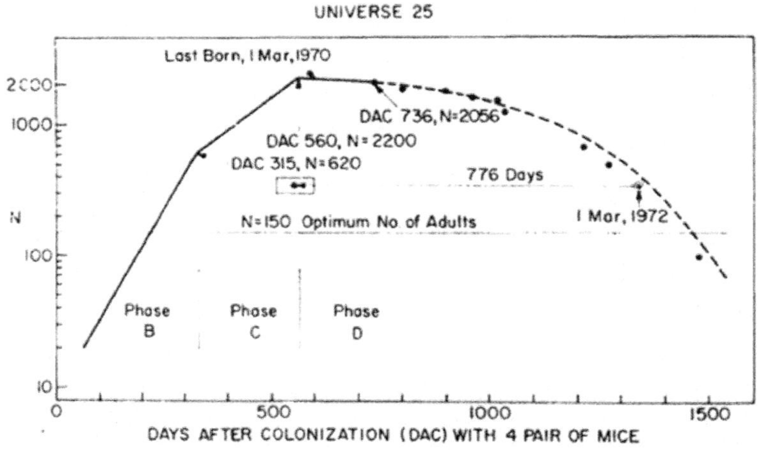

Figure 10 – Calhoun's results from his 1973 paper. Reproduced it for the purpose of discussion and critical analysis.

In terms of quantitative measurements of population size in Calhoun's experiments, the data available in his paper are scant or nonexistent. His 1962 paper,[45] the first of the series, reports no quantitative data on population trends. His main paper, published in 1973,[13] reports only one graph that contains just 14 points. Were more data collected? Maybe, but if so, the results were not made public. Note, in the figure, how the shift from phase B to phase C was determined, apparently, from a single data point, taken almost one year after the start of the experiment.

Many of the statements one can read in the "discussion" section at the end of the 1973 paper are perplexing, to say the least. Upon specific questions, several of the conclusions of the study were justified only on the basis of "Mr. Calhoun believes that…" About the experimental conditions, we learn that "The environment was cleaned, most feces and soiled bedding removed, every six weeks or two months, but nothing was ever sterilized." Calhoun also declared that "the investigators were not very sanitary in their husbandry," and that "Dead bodies were eventually removed for examinations, but the major pollution was the excess of living bodies; this was the essential factor." Maybe. But you might reasonably wonder what the mice thought about the matter.

Considering the density of a mouse fecal pellet, you can calculate the thickness of the excrement layer between cleanings. During the high crowding phase of the experiment, when more than 2000 mice were confined in 6 square meters, the mice would have had to wade through a layer of 3-4 cm of excrement. To say nothing about urine and dead mice. No wonder that the poor critters were not so inclined toward romantic adventures with their fellow prisoners.

How about the possibility that germs may have played a role? It is known that laboratory mice are subject to pathologies that, among other things, can reduce their reproduction rate.[47] The question was asked to Calhoun, and the answer was: "Bacterial culture taken at the highest density of the population indicated that such organisms were not a factor in our study." How many tests were performed? In which conditions? For which pathogens? No answer seems to exist to these questions.

One can't avoid wondering how it was possible that such a poorly characterized experiment and its arbitrary conclusions were so widely praised and had so much success. Today, it is likely that Calhoun's papers wouldn't be accepted for publication in a reputable journal or would face retraction afterward. But so is the way the human mind works; it tends to be attracted by the spectacular and the imaginative. Calhoun was a brilliant scientist, but his brilliance led him to emphasize the spectacular side of his experiments in order to gain the attention of the public. Among other things, he enjoyed engaging in citations from the Bible and in imaginative but uncertain comparisons between the behavior of mice and men. The problem with the reliability of media-oriented scientists is important today, but it already existed in the 1970s.

Yet, we shouldn't throw away the mouse with the bathwater, so to speak. Even though Calhoun's experimental procedure was a little, let's say, 'untidy,' they were similar to those of earlier and later studies. In all cases, the experimenters observed that the population of mice or other animals confined in limited spaces didn't grow forever, as you would have expected. In all cases, the population grew up rapidly at the beginning, then it plateaued, with mice dying of diseases or becoming unable to reproduce. In particular, in the 1960s, Jonathan Terman performed a series of experiments on other rodent species.[48] His animals never completely stopped reproducing, and he didn't observe the population collapse reported by Calhoun. However, he could quantify the idea that crowding affected the reproductive rate of mice by measuring the weight of male and female reproductive organs and comparing the results with those of a control population. Using this correct experimental procedure (that Calhoun never tried), Terman found that the mice raised in crowded conditions had smaller reproductive organs. Other researchers report similar results.

So, Calhoun's interpretation was accurate, despite being based on scant evidence: overcrowding stress does reduce the reproduction rate, at least in some animals. It is a concept that makes sense in evolutionary terms. We don't need to assume that animals are taking rational or conscious decisions about reproducing or not. But individuals who can perceive overcrowding signals in advance can avoid wasting energy on offspring that would have a hard time surviving. The effect of the signal can be indirect, caused mainly by stress through various molecular mechanisms, for which you can find a detailed description in a review by Suvorov. [49] You may also be interested in the review by Viktor Dolnik,[50] who distinguishes between "primary" or "ultimate" factors, which include famine and other Malthusian factors, and "secondary" factors related to indirect effects on the reproductive system of individuals, including stress, pollution, and intraspecific aggression. According to Dolnik:

> "If a species has genetic programs for monitoring changes in signal factors that provide advance warning of increased population density or a decrease in the biological capacity of the habitat, it can stabilize its population or begin to reduce it in advance, before the ultimate factors strike. While control by primary factors is inevitable for any species, only those species in which natural selection has developed special mechanisms for responding to them can take advantage of the warning signals of secondary factors."

In biology, the phenomenon of how populations react to increased crowding is called the density-dependent effect.[51] It is part of the vast field of population dynamics that cuts through different areas of experimental and theoretical biology. Its origins go back to the work of Pierre François Verhulst (1804 – 1849), who started from Malthus's ideas to develop the equation that we call today the logistic equation. Later, in the 1920s, Alfred Lotka and Vito Volterra independently developed the model we nowadays call the "Lotka-Volterra" or "LV" model.[52] These early models remain today at the basis of the field of population dynamics. Details on how they work can be found in the "appendix" of this book, but let's use a qualitative description here.

Malthus' idea was that a population would continue growing geometrically until it was hit by a famine that would brutally stop its growth. It is a concept akin to a car (the population) running into a concrete wall (the population limit) with the driver making no attempt to brake. Verhulst, instead, proposed that the driver would gradually apply the brakes until the car reached the wall at zero speed, gently touching it. Regarding Lotka and Volterra, you have to imagine the wall as an elastic barrier. The car slows down before hitting it, but it still hits it at a certain speed. The wall absorbs the energy by deforming, and then it pushes the car back in the direction it came from. Then, the car accelerates forward again, and the cycle repeats. In the simplest form of the LV model, the oscillations never end.

These considerations hold for theories and controlled experiments, but what can we say about the population behavior outside the lab? Do overcrowding signals avoid the Malthusian famines, at least in some cases? Let's first see an example of a case that has some contact points with Calhoun's experiments, that of the deer of Angel Island in San Francisco Bay.

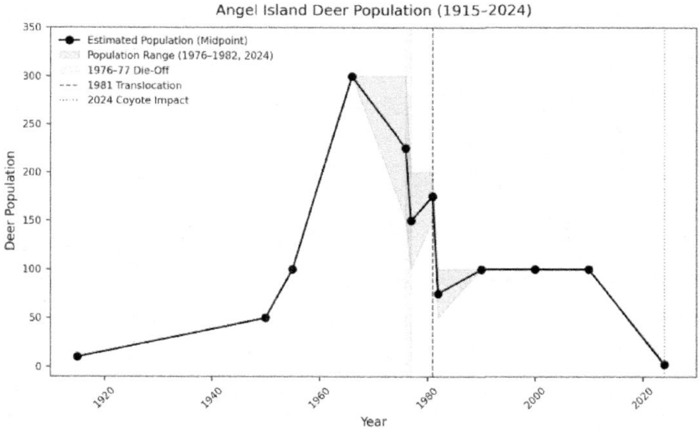

Figure 11 - An approximate reconstruction of the deer population of Angel Island. Data from various sources.

Angel Island is just 8 square km, and it never had a deer population until 1915, when the black-tailed deer was introduced by

Army personnel with the idea of using them as hunting targets. Angel Island became a state park in the 1950s, and hunting was banned in the 1960s. At the peak point, the deer had turned the ground bare by overgrazing, and they were starving and dying. Discussions on what to do ensued in the press, with environmentalists vehemently rejecting the idea of allowing hunting to resume. In 1977, an attempt was made to feed the starving deer with food brought from the mainland, which resulted in a temporary population increase. But that couldn't be a long-term solution. In the end, most deer were relocated to the mainland by helicopters at great expense. Most of them soon died, killed by predators or in road accidents; living on their small island had not prepared them for the challenges of a real ecosystem. Subsequently, the island's population was unofficially stabilized by human hunters. The recent decline visible in the figure was caused by coyotes, which swam to the island and settled there, finding the deer a good source of food.

Did the 300 deer in the 8 square kilometers of Angel Island perceive overcrowding signals? Apparently not, because the population kept increasing until it crashed against the island's carrying capacity, that is, when they started starving to death. Evidently, they didn't behave like Calhoun's mice.

A similar case is that of Miyajima Island, in Hiroshima Bay, Japan. With an area of 30 square kilometers, Miyajima is about 4 times larger than Angel Island, but it had the same problems with deer. It is a renowned tourist spot, but when I visited it in the 1990s, it was not so wonderful to behold. The ground was brown with no traces of grass, and many deer were crowding near the tourists, begging for food. At that time, the deer population is reported to have reached more than 1000 individuals, about the same density as the population peak of Angel Island.

I cannot say whether the Miyajima deer were stressed or not when I saw them. I can only say that the condition of overpopulation was evident: wild deer do not crowd around people, begging for food.

In the 2000s, the Hiroshima city authorities forbade feeding them, and that caused a population crash. It is reported that there are still some 500 deer on Miyajima Island, and they are said to be still overgrazing the land. But there is no evidence that stress caused a decline in the deer reproduction rate.

These cases show that stress doesn't necessarily lead wild creatures to reduce their reproduction rate in order to avoid collapse. Nevertheless, all these cases involved some kind of human perturbation, that is, the introduction of a new species into an environment that previously didn't support them. But none of these cases may be representative of the behavior of populations in the real world.

In a 2005 review in Science,[53] Richard Sibly and his coworkers examined 780 time series of the growth of animal populations in the wild: mammals, birds, fish, and insects. In most cases, the results could be interpreted in terms of the Verhulst equation, that is, with the populations growing at slower and slower rates, until they reach the maximum value allowed by the carrying capacity of the ecosystem. But many populations also showed oscillations. As described by the Lotka-Volterra model, populations tend to "probe" the limits, oscillating over and under the maximum value. In some cases, the slowdown was very gradual, starting much before reaching the population limit. In others, it started only when it was close to the limit. But, in general, and in particular for mammals, the slowdown started before the limit was reached. Evidently, most wild creatures can perceive overcrowding signals and act accordingly.

What Sibly's review cannot tell us is how exactly the slowdown occurs. Is it caused by reduced food availability? Or by overcrowding stress? Or maybe by some other factor? Field studies tend to measure population trends in terms of the number of individuals, but rarely provide specific data on birth and death rates. The problem is typical of complex systems: when several factors are pushing the system in the same direction, it is extremely difficult to disentangle the respective contributions. The best we can say is that animal populations in the wild tend to stabilize their numbers near the carrying capacity of their environment, provided that they are not overstressed by external factors.

What about human beings? Could they be sensitive to stress generated by overcrowding signals? Could that entirely explain the current decline in the global birth rates all over the world? After all, nowadays we are not seeing large primary Malthusian factors at work: famines, pandemics, and mass exterminations exist but, at present, they are affecting only a relatively small fraction of the world's population. Even the much-hyped COVID-19 pandemic was short-lived and didn't cause a significant decline in the global population. So, could it be that the slowdown we observe is mainly caused by stress, in turn generated by overcrowding?

In principle, it is possible. The medical literature tells us that stress negatively affects human health, including the reproductive system. In women, higher stress levels are associated with irregular menstrual cycles, anovulation (lack of ovulation), and oligomenorrhea (infrequent menstrual periods). In men, stress has been linked to poor sperm quality, including lower sperm count, reduced sperm motility, and higher levels of sperm DNA damage. For both sexes, stress can alter the levels of reproductive hormones such as estrogen and progesterone in women and testosterone in men.

These hormonal changes may mean that people lose interest in sex, something not unlike what happened to Calhoun's mice, although humans are generally not forced to wade through a thick layer of their excrement in order to have sex, as Calhoun's mice were. But is pure stress a significant factor in current trends? As I said before, it is almost impossible to disentangle the role of single factors pushing in the same direction in complex systems. On this point, you may read *Stress R Us* (2018) by Greeley G. Miklashek to gain some feeling on the complexity of the matter. Nevertheless, we can try to orient ourselves by looking at the available data sets.

So, first of all, how do we define "stress"? In the medical literature, you can read that stress is "a physical, mental, or emotional factor that causes bodily or mental tension." That's a little generic, to say the least. If you look for quantitative measurements, you find that the main tool used today is self-reporting, and we may suspect that these surveys may not be so reliable. Different people in different cultural and social environments may perceive stress in widely different ways.

More quantitatively, stress can be detected by the symptoms it generates. In humans, they appear in terms of disrupted sleep patterns, anxiety, irritability, and the like. In animals, they may appear in the form of increased aggressiveness, lack of interest in social interactions, and loss of interest in sex. But if stress is a thing, then there have to be methods to measure it in terms of actual changes in the way the body operates. It can be done by measuring the cortisol level, also called the stress hormone. Cortisol is a steroid, one of those molecules that play a crucial role in many processes inside the body. Its main role is to respond to threats by raising blood sugar levels, providing an immediate energy source for the "fight-or-flight" response. In addition, it regulates inflammation, blood pressure, the metabolism of fats, and probably more metabolic cycles. Unfortunately, cortisol is not normally measured in sane human beings, so we don't have large-scale data that could allow us to correlate how stress varies in time with factors such as crowding. The best we can do is to use "proxies," that is, trying to correlate the changes in fertility rates with parameters that are known, or at least suspected, to increase stress.

If stress is a cause of infertility, we should observe more stress in urban settings than in rural ones. Generally speaking, it appears to be true, as one can read in a review by Peen et al., [54] even though the difference in stress levels is not very large. But is higher stress in cities the only factor that generates infertility? The figure shows the worldwide total fertility ratio (TFR, number of children per woman) as a function of urbanization. Similar results were found in a recent post by Anastassia Makarieva.[55] The world's fertility rate took a tremendous plunge starting in the 1960s, but what caused it?

The figure shows a correlation, but, of course, correlation doesn't imply causation. Did humankind hit some kind of threshold when the global urbanization fraction reached about 35% of the total population? It cannot be excluded, but why should that have occurred? If stress is related to population density, it is hard to maintain that the world's cities in the 1960s and 1970s were so much denser than those of the 1950s.

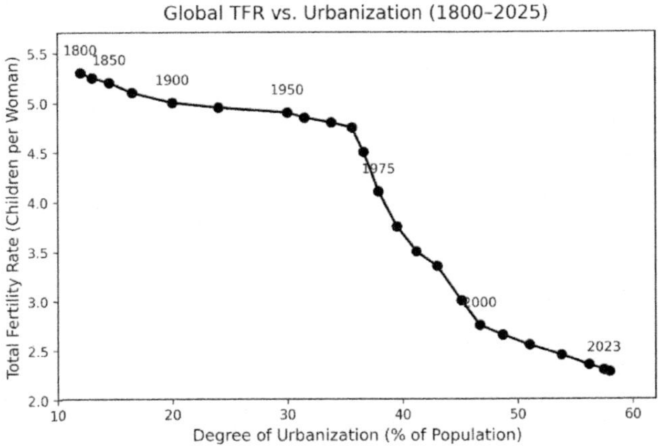

Figure 12 -- Global urbanization (% people living in urban areas). Data from "Our World in Data"

Taking a different approach, we may examine the case of the United States, the paradigmatic Western nation. We have data from the National Health Institute (NIH)[56] telling us that in the 1950s, about 10%-15%of the US population reported that they were suffering from high levels of stress.[57] In 2022, about 25% of the population reported the same thing and that fraction changed very little over different age ranges. Assuming that these numbers correlate to actual physical conditions (not at all obvious), it is clear that people are more stressed now than 50 years ago. But is that enough to justify the TFR decline? If 25% of the women in their reproductive age declared that they are heavily stressed, that doesn't mean they had zero children. And even if that were the case, it is not enough to justify the fact that the fertility rate in the US went down from 3.7 in the 1950s to the current value of 1.6, when the difference in the number of stressed and not stressed women was only about 10%.

We may also examine a recent study by Gallup,[58] which reports the results of a poll conducted in 121 countries. It asked questions about positive and negative feelings, which may be considered a qualitative measure of stress. If the results are plotted as a function of the TFR for the respective countries, the result is a weak dependence in the direction opposite to the one expected. That is, highly stressed populations tend to have more children per woman than less stressed ones. Even in some areas of the world where the living conditions are far from being optimal, the human fertility rate doesn't seem to be affected. People living in the crowded suburbs surrounding modern metropolises seem to have higher fertility rates than people living in apartments. As an example, the Italian Roma (Gypsies) tend to live in crowded mobile homes or barracks in open-air camps. Still, their fertility rate is higher than the average in Italy. As another example, the fertility rate of people living in South African shantytowns around the main cities was estimated in 2016 by the South African Demographic and Health Survey (SADHS) as about 3.3 children per woman, larger than the 2.3 average of South Africa. Evidently, humans can adapt to conditions of overcrowding which would make a mouse's brain explode.

In general, sex and reproduction are fine-tuned machinery governing the behavior of biological species; it is certainly true for humans as well as for mice. They may well be affected by psychological factors related to stress generated by crowding, but from the available data, we can only arrive at tentative and uncertain conclusions. Although stress factors may play a role in human behavior, they are insufficient to explain the abrupt decline in fertility rates observed starting in the second half of the 20th century. Other factors, chemical, social, and economic, must be taken into consideration, as we'll see in other sections of this book.

The evolution of the social system has led us to where we stand, highly stressed by many different factors related to crowded spaces: noise, criminality, government propaganda, debts and financial problems, family problems and obligations, and more. Perhaps we will undergo the same fate as Calhoun mice: so stressed that they lost interest in reproduction and went extinct. But the society in which we live changes continuously and evolves, and we don't know if the future has more or less stress in store for us.

Carson: Pollution

Rachel Carson (1907-1964) originated some fundamental concepts that are still part of our worldview today. She popularized the concept of "chemical pollution," and her book *Silent Spring* (1962) was a milestone; the start of what we call the environmental movement.

As is customary in our times, whoever proposes a new idea is subjected to a barrage of insults and demonization, especially by those people whose economic interests are harmed by the new views. Carson's work was no exception, and the lobby of the chemical industry mounted a strong attack on her. She was said to be a fanatic, little more than an enthusiastic bird watcher, and a Communist to boot. She was also termed a "Priestess of Nature," which was supposed to be an insult. I don't know what a priestess of Nature should look like, but I think the term fits her perfectly as an avatar of the Goddess Gaia. The attacks didn't succeed in forcing Rachel Carson to silence, and several high-profile disasters generated by pollution strengthened her arguments.

Rachel Carson didn't deal specifically with the human population trends. Still, she mentioned the loss of fertility of animal species due to pollution by synthetic pesticides and their widespread, often indiscriminate use. For example, she discussed how DDT and related pesticides interfere with reproduction in birds, leading to thinned eggshells and reduced hatching success—a phenomenon later linked to the endocrine-disrupting properties of pesticides. In Chapter 8 of her book, titled "And No Birds Sing," she described how robins and other birds exposed to DDT showed population declines partly due to reproductive failure. Carson also wrote about how synthetic chemicals could accumulate in human tissues and hinted at broader implications, including possible links to reproductive harm. She wrote about how these substances "pass from mother to unborn child" and disrupt biological processes, though she didn't claim that they could lead to widespread human infertility.

At the time of Rachel Carson, little was known about the biology at the basis of human fertility trends. Most people reasoned on the basis of the Malthusian concept of the urge to reproduce. They believed that the human population would keep growing rapidly in the foreseeable future, skyrocketing toward disaster. But there were already hints that something ominous was brewing in the human reproductive system, just like in that of birds. Starting in the late 1930s, John MacLeod was among the first researchers to analyze semen quality in American men, establishing a normal average level of ~100 million/milliliter (mL).[59] He didn't note any decline in sperm count, but that would soon become evident in later studies.

The "sperm count decline hypothesis" was proposed in the 1970s, although the initial findings were inconsistent. The turning point came with a paper by Carlsen et al, published in the British Medical Journal in 1992.[60] Reviewing 61 studies from 1938 to 1990, the authors found a 50% decline in sperm concentration among men in

Western countries over five decades. Later, Shanna H. Swan's research refined the 1992 findings. Her 2017 meta-analysis [12] covered 185 studies and 42,000 men from 1973–2011, confirming a 50–60% drop in sperm concentration in North America, Europe, and Australia/New Zealand. This result went in parallel with the data for the global Total Fertility Rate (TFR). Swan popularized her results in her book *Countdown*, published in 2022.

The World Health Organization (WHO) has been monitoring this decline,[61] opting to do little more than reduce the minimum value of sperm concentration that guarantees fertility for human males. Now it is about 15 million/ml, and even so, it is reported that 17% of the conscripts in the Swiss Army had sperm levels below the fertility level in 2019.[62] If we assume that the average sperm count continues to decline at 2.6% yearly, as it is doing now, it follows that more than 50% of Western males will go below the limit in less than 50 years. And it could happen much faster since the rate of decline in sperm count appears to be increasing (another manifestation of the Seneca Effect?). If that happens, it doesn't mean that no children will be born, but that by the second half of the 21st century, a majority of human males will be infertile. Today, males in African countries have a larger sperm count than in other regions of the world. Still, the declining trend operates for them, too, and African males could become mostly infertile well before the end of the century.

The sperm count is not the only physiological factor affecting fecundity. Another is sperm motility and morphology, which appear to go in the same direction as the sperm count. And even a healthy sperm cell can carry damaged DNA with it. Other factors, not directly linked to sperm cells, include hormonal regulation and reproductive organ function. Then, data about the loss of testosterone among men exist, indicating a lowered sexual desire, and, according to some researchers, a general "feminization" of males.[63]

Infertility is not just related to male dysfunctions. Women have their problems, too. The number and quality of the egg cells normally decline with age, but environmental factors (e.g., dioxins) may accelerate the effect. Exposure to pollutants can cause mitochondrial dysfunction or chromosomal abnormalities in oocytes, lowering fertilization success and embryo viability. In general, there is plenty of evidence of pollution-related physiological decline in both human males and females, but also in many other mammals and birds that share similar metabolic and endocrine mechanisms.

The main culprits in this field are several substances contained in plastics and released into the environment, or plastics themselves when dispersed in the form of very small particles, at the scale of one millionth of a meter, micron (μm), or one billionth of a meter, nanometer (nm). Plastic waste is often considered an inconvenience, something that's ugly to see but not really dangerous. But plastics are not inert. They are carbon-based polymers made by assembling smaller molecules, "monomers," to form molecular chains ("polymers"). The result is a stable solid. However, the solids slowly degrade, releasing the chemicals they contain and forming tiny particles of microplastic, which end up dispersed in the environment. These chemicals enter the food chain and are eventually ingested by human beings through food and water. They are especially common in foods such as seafood, salt, and honey, but they can be found in every kind of food, as well as in tap water. It appears that every one of us has an amount of plastic microparticles equivalent to a plastic supermarket bag or a credit card in the brain.[64] We don't exactly know what kind of problems it could lead to, but it is hard to think that it would make us smarter and healthier.

The body may treat microplastics as foreign invaders, triggering chronic inflammation or immune dysregulation. Chronic inflammation and DNA damage from microplastics could also elevate cancer risk. Recent data show that microplastics in the human body are associated with chronic non-communicable diseases (NCDs), including hypertension, diabetes, and stroke.[65] Most of these problems do not directly affect reproductive health, but sick people can hardly think about having children.

Microplastics may turn out to be a relatively minor problem in comparison to the products of their decomposition, slowly released from the particles. According to UNEP,[66] more than 13,000 additives are used in plastics, some of which are not added intentionally, and we don't even know where they come from. Testing for the health effects of 13,000 chemicals is simply impossible, to say nothing about testing their combined effects. We can only be certain that we are all exposed to potentially harmful substances produced by plastics when they degrade. These substances don't kill people right away, but what about long-term effects?

Plastics are known to contain several substances that alter the endocrine system of living creatures, including human beings. The list is long and based on arcane names such as organochlorine pesticides, polychlorinated biphenyls (PCBs), dioxins, furans, bisphenol A (BPA), polyfluoroalkyl substances (PFAS), and many others. Most of these substances are dispersed in the environment from the degradation of plastics, from their manufacturing, or other chemical and mechanical processes.

As an example, a major pollution disaster involving polyfluoroalkyl substances (PFAS) recently took place around the town of Trissino near Vicenza, Italy. It was the result of years of mismanagement of the Miteni chemical plant in Trissino that released PFAS in the wastewater for decades, starting in the 1960s. The story came to public attention in 2013 when a study commissioned by the Italian Ministry for the Environment revealed massive PFAS pollution in the drinking water, soil, and rivers of the Italian Veneto region. The Italian National Research Council's Institute of Water Research found high levels of PFAS compounds across a 200-square-kilometer area. An estimated 350,000 residents had been exposed. The whole food chain was affected—livestock, crops, and fish carried PFAS, leading to bans on locally produced food and well water use.

Figure 13 - The "mamme no PFAS" of Trissino. Reproduced from mammenopfas.org

The Italian National Institute of Health conducted blood tests on the residents in 2016-2017. The results indicated PFAS levels up to 20 times higher among the residents of the area than among the unexposed populations. The "acceptable" amount of PFAS in the body is four ng/ml (nanograms/milliliter). In the contaminated areas near the Miteni plant, values of 20–100 ng/mL or even higher were commonly found in the local residents. Plant workers had levels in the thousands.

The chemicals of the PFAS family are known as "forever chemicals." They are extremely unreactive and nearly impossible to get rid of, and remediating the site is a nightmare in terms of cost, even assuming that it will be possible. Activated carbon filters can treat drinking water, but the poison is everywhere. There was little that the residents of the area, or the government, could do. The best they could manage was to form a protest group called "Mamme No PFAS" (Mothers Against PFAS), who would wear in protest T-shirts with the PFAS value found in their blood.

The PFAS problem in Trissino was not just numbers printed on t-shirts. A 2018 study[67] found higher death rates for diabetes, cerebrovascular diseases, and myocardial infarction in contaminated areas compared to similar uncontaminated ones. And it did affect fertility: a 2019 article in the Journal of Clinical Endocrinology and

Metabolism[68] shows significant reproductive health impacts in the reproductive organs among those exposed to PFAS. Specifically, the study found reduced testicular volume, shorter penile length, and a lower anogenital distance (an indicator of prenatal testosterone exposure) in exposed individuals compared to the control group. Additionally, there was a noted decrease in fertility potential, with sperm counts reduced. The Miteni company filed for bankruptcy in 2018, and its managers now face criminal charges. But even if they are condemned to monetary compensation, the residents won't get their health back.

The story of the town of Trissino is just an example of events happening everywhere in the world with PFAS, but also with other chemicals. Consider bisphenol-A (BPA). It is a small molecule used to create polycarbonate chains, one of the most common forms of plastic. It is also used to make epoxy resins, vinyl resins, and is a component of the thermal paper used in cash registers and other thermal printers. You find it everywhere, not just in plastics but also in coatings for the inside of food cans, clothing, shop receipts, and dental fillings.

BPA is known to be a "xenoestrogen." That is, it has properties that mimic the effects of estrogens in the human body. Estrogen is one of the fundamental hormones in the endocrine system, more concentrated in females than in males. In women, it controls several sexual body characteristics. In men, excess estrogen (or xenoestrogen) will cause a fat belly, large breasts, and lowered sex drive. In addition, for both males and females, it may increase the risk of cardiovascular disease, type 2 diabetes, high blood pressure, and infertility.

So, the world's infertility epidemic is what you would expect if endocrine-disrupting additives, such as BPA, PFAS, and others, were having their known effects on the entire population. Unfortunately, we could prove it only by phasing out BPA and PFAS and then, years later, see the fertility trend reversing. But do we need proof to suspect that these endocrine disruptors and xenoestrogens are wreaking havoc with people's delicate and fine-tuned endocrine systems?

Wouldn't it be a good idea to phase it out, or at least curb its production, as soon as possible?

Nevertheless, curbing plastic production is anathema to the current way of thinking when the accepted paradigm is that growth is the perfect thing. Producers are proudly rolling out "BPA-free" plastics. Nice, but these products typically contain other forms of bisphenols, such as BPS or BPF, which were found to have the same harmful effects as BPA.[69] This is not surprising: they are very similar molecules containing aromatic rings. Why should they be safer? Fortunately, there are now true "Bisphenol Free" plastics in the market, but they are arriving late, and the market is still dominated by the harmful kinds.

The story of how chemical pollution affects human fertility could go on, considering materials other than organic molecules related to plastics. Another sector, probably just as important, is the effect of heavy metals. The list is long, and it includes such metals as lead, mercury, cadmium, arsenic, and many others. Sometimes, metals are dispersed in the environment with the purpose of fighting a kind of pollution, without paying attention to the side effects. A good example is that of automotive catalytic converters, which use noble metals, palladium, rhodium, and platinum, to operate. These metals are extremely rare in nature, but their massive introduction in heavily populated areas exposes human beings to them as a result of the slow degradation of catalytic converters. It is an especially dangerous experiment carried out on the health of all humankind, because we have no idea about the long-term effects of metals to which the human metabolism was never exposed.

Heavy metal pollution affects human health and fertility primarily by causing oxidative stress, disrupting hormonal balance, and damaging cells and DNA. Heavy metals accumulate in vital organs. For instance, cadmium can cause kidney and bone damage, while mercury and lead are known to harm the brain and nervous system. Lead exposure in children can lead to neurodevelopmental and behavioral issues. Many heavy metals are considered carcinogens, increasing the risk of cancers in the lungs, liver, and skin.

The effects of heavy metals on human fertility are known: they interfere with cellular functions by binding to proteins, displacing essential metal ions, and generating free radicals that cause cell damage. These metals can act as endocrine disruptors, interfering with hormones essential for reproduction, such as testosterone and estrogen. This can cause imbalances that lead to infertility. In men, exposure to heavy metals has been linked to reduced sperm count, motility, and viability, as well as damage to sperm DNA. In women, heavy metal toxicity can contribute to decreased egg quality, menstrual disorders, endometriosis, and an increased risk of miscarriage, stillbirth, and birth defects.

To these chemical effects, we should add other forms of pollution, which include electromagnetic pollution and radioactivity. Research indicates that exposure to non-ionizing electromagnetic fields (EMFs), such as those from mobile phones, Wi-Fi, and electronic devices, negatively impacts reproductive health. Studies have found associations between EMF exposure and reduced sperm quality, including decreased sperm count, motility, and viability. The proposed mechanisms include the induction of oxidative stress, which can damage sperm DNA and cellular membranes. The testes are particularly vulnerable to this type of radiation. In animal studies on females, EMF exposure has been shown to disrupt the estrous cycle of females, alter hormone levels, and cause a decrease in the number of ovarian follicles. These effects can lead to ovarian dysfunction and a higher risk of infertility.

Then, there is radioactivity: ionizing radiation is a well-established reproductive hazard. Its effects are highly dependent on the dose, duration of exposure, and the age and gender of the person exposed. Male testes are highly sensitive to radiation. Doses as low as 0.15–0.5 Gy can reduce sperm count, and higher doses can lead to temporary or permanent sterility. Radiation can damage sperm-producing cells and cause DNA mutations that can be passed on to offspring. In females, ionizing radiation leads to premature ovarian failure and early menopause. Additionally, radiation exposure during pregnancy can increase the risk of miscarriage, stillbirth, and birth defects.

Other forms of non-chemical pollution include noise, light, and high temperatures. Chronic exposure to loud noise can activate the body's stress response system, leading to elevated cortisol levels, which can disrupt the delicate hormonal balance required for reproduction in both men and women. In men, noise has been linked to a decrease in testosterone levels and a decline in sperm count, motility, and viability. In women, it can disrupt menstrual cycles and ovulation, and has been associated with an increased risk of miscarriage.

Regarding light pollution, artificial light at night can interfere with the body's circadian rhythm, which is a key regulator of reproductive hormones. The pineal gland, which synthesizes the sleep hormone melatonin, is particularly sensitive to light. When melatonin production is suppressed by light at night, it can lead to hormonal imbalances that may impair fertility in both sexes.

Finally, thermal pollution coming from global warming is known to negatively affect male fertility. The human testes require a temperature that is a few degrees cooler than the rest of the body for optimal sperm production. Exposure to high heat can lead to reduced sperm count and motility. Studies have also shown that global temperature increases are correlated with a decline in birth rates.

The general phenomenon of pollution affecting fertility is well known and documented. Yet, often the discussion on the human fertility decline tends to ignore it, preferring to attribute the decline to socioeconomic factors. It is possible that this preference is due to the realization that acting on socioeconomic factors is possible with limited efforts, such as subsidies and taxes. Instead, acting on pollution requires going against the way industrial society works, typically requiring reducing profits by dedicating money and resources to reduce pollution. Unfortunately, the world doesn't work according to human preferences, and in the future, we will have to address the pollution effect on fertility as a prerequisite for the very survival of humankind.

5. Population Control

What checks the natural tendency of each species to increase in number is most obscure. Charles Darwin, "On the Origin of Species" (1859)

Figure 14 -- The young widow Tamar seduces her father-in-law, Judas, according to the story told in the Genesis book of the Bible. Painted by Rembrandt Van Rijn's school, 1650-1660. Wikipedia, public domain.

In the previous sections, we saw the external factors that affect population, and limit it to below a maximum possible level. In this section, instead, we will see how humans themselves act on population; a field called "population control." We will start from the earliest times, when birth control was officially ignored and often demonized, to our times, when it is debated among "catastrophists," those worried about overpopulation, and "natalists," those who think that an increasing population is a good thing.

Cannon Fodder

The Bible is one of the earliest texts that mentions population control, although in the opposite sense as it is understood today. In the Book of Genesis, possibly written in part in the 10th century BCE, you can read the sentence "Be fruitful and multiply, and fill the earth."

This Biblical command is a special moment in the history of population control. For most of humankind's existence, having children was a question decided by families, and, in particular, by women. Even without modern contraceptive methods, throughout history, women knew how to manage their pregnancies by abstinence, or, when necessary, by abortion or infanticide. But a command coming from a higher authority implied that women were no longer allowed to choose how many children they would have. Most likely, the Jews emphasized population growth because they were living in an area that was a crossroad between Asia and Africa, continuously contested among different peoples living in the nearby regions (it still is).

Apart from the Bible, only a few ancient thinkers and prophets wrote about these matters, and it doesn't appear that any of them proposed such an unqualified exhortation as "multiply and fill the earth." Aristotle (4th century BCE) noted how the population of a city-state had to be managed to maintain a certain optimal level. In Asia, Confucian thinkers Han Fei (280-233 BCE) and Xunzi (ca. 310-235 BCE) proposed that it was the duty of the central authority to manage society, including regulating the population size according to the available resources. Roman Emperors understood the need of citizens in order to replenish the depleted legions of the late Empire, and some of them enacted laws to favor large families. Still, it doesn't seem that having children was ever considered a sacred duty in ancient Rome. In Islam, there are no explicit statements about population in the holy Qur'an, but, in line with the generally moderate tone of Islamic teachings, believers are advised to have as many children as they can reasonably provide for.

Despite the injunction "be fruitful and multiply," early Christian thinking carried a strong emphasis on chastity and on the view that sex itself was something evil. Christian views were not unique in Eurasian thought. Buddhism, too, praised chastity as a virtue for the wise. Neo-Platonists emphasized transcending bodily desires, and Jewish ascetic communities like the "Therapeutae" practiced celibacy. But the views related to Christianity were extreme. An early Christian father, Origen of Alexandria (185-254 CE), is said to have castrated himself to avoid all temptations and, although Origen's act was later condemned by the Christian Church, it illustrates the Christian attitudes of the time. The Gnostics, a series of religious movements that emerged at the beginning of Christianity, were particularly rigorous in their views on chastity. In early Christian and Medieval times, a general attitude prevailed that regarded chastity as the primary virtue for both men and women. If you read documents such as the *Pratum Spirituale* (The Spiritual Meadow) by John Moschos, written in the late sixth to early seventh century, you get the impression that people at that time were obsessed with sex, that is, how to avoid it.

The exaltation of chastity in Christianity may have been the result of a time of economic contraction that led to an attempt to avoid rapid population growth. However, with the waning of the Middle Ages, Europe saw a phase of robust economic growth driven by commerce with Mediterranean countries and, later, by expansion in the New World. In parallel, Europe saw a decline in attitudes toward chastity, with soldiers and colonists needed to expand overseas.

During this historical phase, Catholic priests remained officially chaste in the economic backwaters of Southern Europe. Instead, the more dynamic economy of Northern Europe would cause the decline of chaste monks and nuns and allow priests to marry and have children. Martin Luther married a nun, Katharina von Bora, one of 12 nuns he had helped escape from their Cistercian convent in April 1523. Note that Katharina was part of a noble family, the Haugwits of Saxony, as it was typical to have daughters "cloistered" to avoid having to pay a large dowry for their marriage. Among other features of this period, it is said that the witch hunts of the 16th century were the result of rulers attempting to stop sage-women and midwives from helping women in how to avoid unwanted pregnancies, and also in how to interrupt them. We have no way to prove that it was the case, but this story may have some elements of truth.

The attempt to turn families into soldier-manufacturing machines had its zenith during the 19th and early 20th centuries. Numbers had always been important in battle, but in many cases, better-trained and better-armed armies could defeat larger armies of less experienced and less well-equipped fighters. But with the development of firearms, battlefields became large-scale versions of butcher shops. Musket-armed infantry on both sides marched onward while being torn to pieces by bullets and grapeshot. The carnage was appalling, and heroism and competence had little importance. Consequently, one rule emerged: those who had the larger armies usually won.

The importance of large armies led modern leaders to push for natalist policies. At the same time, if people had to agree to march onward on the battlefield while being sprayed with bullets and artillery shells, they had to have at least something in return. Governments had to provide health care services, retirement pay, and a certain financial stability for families. Above all, soldiers would stay away from home for years, and governments enacted laws that promoted nuclear families, ensuring that the Biblical command "Thou shalt not covet thy neighbor's wife" was enforced by laws and by the police.

For most of the past four centuries, the dominant view among rulers and states was that a large and growing population was a direct source of national wealth, military power, and fiscal revenue.

Mercantilist doctrine (roughly 1650–1800) explicitly equated "population with power," leading European states to adopt aggressive pronatalist measures.

France was probably the pioneer of these policies. One of the first countries that saw fertility decline, it also saw leaders trying to counter the trend. As early as 1666, Louis XIV offered bounties for large families and granted pensions to fathers of ten or more children. The Third Republic (1870–1940) introduced the most elaborate system for natality encouragement in Europe: family voting (extra votes for fathers), tax exemptions scaled by number of children, medals for "*mères de familles nombreuses,*" and heavy taxation on bachelors. The 1920–1939 *Code de la Famille* provided generous family allowances paid by employers, housing priorities, and birth premiums — widely credited with raising French fertility from 1.7 in 1935 to almost 3.0 during the immediate post-WWII "baby boom."

Fascist Italy and Nazi Germany took pronatalism to its most extreme form in the interwar period. It was typical of dictatorships, where the great leader often took the role of "Father of the Country." The Italian Duce, Benito Mussolini, was the prototypical figure of the alpha male in politics. The Fascist government (1922-1943) generated plenty of natalist propaganda and offered incentives and benefits for large families. The slogan "Eight Million Bayonets" has remained a proverbial example of government stupidity, having been expressed in an era when bayonets were already obsolete weapons. In 1927, the "Battle for Births" came with a bachelor tax, gave employment preference to married men with children, awarded medals and cash for high fertility, and criminalized abortion and contraception propaganda.

Hitler's regime went further: the Mother's Cross (gold for 8+ children), massive marriage loans forgiven at 25 % per child. The Lebensborn (fount of life) was designed to be the wellspring of future generations descended from those whom Nazi authorities deemed "racially valuable," typically "Aryan." Both regimes saw fertility rise temporarily (Italy from 23 to 30 births per 1,000 in 1925–1935; Germany from 14.7 in 1933 to 20.4 in 1939), though the gains collapsed after the war.

The Soviet Union was not pro-natalist at the beginning, but it turned pronatalist when labor or soldiers were needed. Stalin banned abortion in 1936, introduced "mother-heroine" titles for ten children, and paid large stipends per child. After the catastrophic population losses of World War II, similar measures were reintroduced in the 1940s and again in the 1980s under Brezhnev and Gorbachev.

The British Empire had a mixed approach, although it was active in promoting population growth. From the 18th century, the Poor Laws and Settlement Acts discouraged the emigration of the able-bodied, but colonial authorities recruited families with promises of land. When the Great Famine struck Ireland in 1846, the British authorities did nearly nothing to avoid the loss of nearly half of the Irish population; evidently considering that it was already large enough. After WWII, Britain, Canada, Australia, and New Zealand all introduced universal "child endowment" or "family allowance" payments (1945–1946) explicitly to reverse low depression-era fertility and rebuild population after war losses.

In Northern Europe, Sweden (1930s onward) pioneered the modern welfare-state approach. The idea was to use tax relief, subsidized housing, maternity leave, and child allowances to favor childbirth. These policies were associated with a strong increase in births from the late 1930s through the 1950s.

In the rest of the world, and in particular in the Third World Countries, pronatalism persisted into the late 20th century in places that still saw themselves as underpopulated. Israel's government has been supporting population growth since 1948 (birth prizes, child allowances rising with parity, free IVF). Singapore under Lee Kuan Yew initially encouraged larger families among university graduates (1984 Graduate Mothers Scheme) before famously reversing course.

The case of China, first natalist, then strongly involved with birth control, is discussed in a specific chapter in this book.

Independent India had an oscillating behavior. Strong anti-natalist policies were initially favored, but for the first 15–20 years after 1947, significant voices inside and outside government actively pushed in the opposite direction – advocating pronatalist measures to build a large population for national strength, military power, and industrial labor. The 1961 Census slogan was openly celebratory: "A population of 439 million – a mark of our strength." By the mid-1960s, however, the balance decisively shifted. Successive droughts and food crises made aggressive anti-natalism the official policy. Sterilization camps, targets, and outright coercion became the dominant approach. Nevertheless, the Indian population continued to grow, reaching one and a half billion people in recent times. With a fertility rate now below the replacement level, the government is trying to encourage natality.

In most of the world, the phase of encouragement of natality lasted from the 17th century up to recent times, with governments repeatedly using cash payments, tax policy, medals, housing privileges, employment preferences, and legal coercion to raise birth rates whenever rulers believed population growth served state power, economic expansion, or post-war reconstruction. This phase exhausted itself from the 1970s onward, when most developed countries abandoned pronatalism in favor of neutrality or open anti-natalist policies. Today, a new phase of pronatalism seems to be starting, but we have to see how it will evolve.

Racism, Social Darwinism, and Eugenics

Among the many original ideas that the Reverend Thomas Malthus proposed, one was that famines and plagues could be avoided by human action, that is, by birth control. It was a very innovative idea. In earlier times, the idea of family planning was concentrated on the opposite goal: how to have as many children as possible. That Malthus proposed this idea is all the more remarkable if you consider that he was a Christian minister, and that the general Christian views of the time were that children were a gift from God, and that humans had no say on how many they would receive.

Of course, Malthus was also a man of his times, and he wouldn't even dream of proposing that good Christians should use contraceptives – that would have gone directly against the Biblical command of "grow and multiply." Rather, he proposed that married couples should practice chastity, a behavior that Christianity had always seen in a favorable light. That was a way to kill two birds with one stone: make God happy by avoiding the sin of sex, and at the same time, avoid overpopulation and hence famines.

As you may imagine, Malthus's call for chastity was never very successful. Still, the idea created a banquet of consequences. What if people could decide not just how many children they would have, but also with whom to have them? And who had the power to make these decisions? It is a problem that Malthus probably couldn't have imagined, but which emerged after him, and we still have to solve it.

Malthus' ideas deeply influenced Charles Darwin in several ways, including the competition for resources and the geometric growth of populations. Both Malthus' and Darwin's ideas were badly misunderstood, but Darwin's theory was the one that had the most nefarious effect on politics. Herbert Spencer (1820-1903) coined the phrase "survival of the fittest" and applied Darwin's ideas to social, political, and economic issues, suggesting that social classes and races competed for survival and only the "fittest" should thrive.

These concepts didn't exist in Darwin's theory. It is correct to say that Darwin's ideas imply competition, but our current understanding of evolution does not imply a "battle royale" of everyone against everyone else. That's true for most living species, and in particular, humans are selected mostly for their capability to get along well with their fellow humans, providing help and being helped when they need it.

Unfortunately, the idea of evolution as a bloody affair, "nature red in tooth and claw," as Tennyson said, soon spilled over to speculations about whether some human races were "fitter" than others. Typically, that led to a pseudo-scientific hierarchy that saw Whites, or "Aryans," being at the top, with the "Darker Races" at the bottom, just a little above monkeys and apes. Of course, Darwin never said anything like that, and, besides, the idea of the inferiority of non-European races was characteristic of European colonial expansion well before Darwin. Nevertheless, this misinterpretation of Darwin's ideas provided a further justification for the colonial practices of slavery and segregation.

These concepts gradually percolated into the human sphere of beliefs, and by the late 19th and early 20th centuries, they led to the development of a set of ideas that went under the name of "Social Darwinism." In a crooked sequence of logic, some people understood that their survival in the fight for existence implied that they should enslave or exterminate other humans, preferably those who didn't belong to their ethnic and linguistic group. Social Darwinism was the first step toward the development and implementation of "eugenics" ("good generation"). The idea was to improve the genetic composition of human beings by selective breeding. In particular, the idea was to remove those components of humankind that were regarded as inferior.

The eugenics movement was influenced by the works of Francis Galton (1822-1911), a cousin of Charles Darwin. He hadn't really understood his cousin's theories, but he used his partial understanding to coin the term eugenics in 1883. His ideas were further popularized by figures like Charles Davenport (1866-1944), who established the Eugenics Record Office in the United States in 1910. This organization collected data on family histories to identify hereditary traits and advocated for policies such as sterilization and immigration restrictions to prevent the spread of "undesirable" genes. Another proponent of social Darwinism was William Graham Sumner (1840 – 1910), a sociologist at Yale. He argued that social hierarchies were natural and that attempts to improve the conditions of the lower classes were futile. Benjamin K. Hays (1887-1918) attempted to use a form of Social Darwinism to rank human races, suggesting that some were more evolved than others.

These ideas were popular in the United States even before Darwin proposed his theory, but they went by different names. A typical one was the concept of "Manifest Destiny," which stated that the expansion of the Whites across the continent was inevitable and divinely ordained. This belief was combined with a sense of cultural and racial superiority among white settlers. It was used to justify the displacement and the extermination of the Native Americans of the continent.

In the United States, these beliefs went in parallel with anti-miscegenation laws that prohibited interracial marriage. These laws were in place for over three centuries, dating back to the colonial period. The first case of a law forbidding mixed-race marriages was enacted in Maryland in 1661, and Virginia followed suit soon after. By the time of the Civil War, many states had such laws that continued to proliferate in the late 19th and early 20th centuries. The motivations behind these laws varied but were deeply rooted in maintaining a racial hierarchy and upholding white supremacy. Some argued that interracial marriage was against the "laws of nature" or that it would lead to social and moral degradation. The laws often extended beyond black-white unions to include other racial groups, such as Native Americans, Chinese, and Japanese.

Stronger eugenic measures began to be practiced in the US around the beginning of the 20th century. Psychologists and researchers were involved in efforts such as the Committee to Study and Report on the Best Practical Means of Cutting Off the Defective Germ-Plasm in the American Population. In 1914, the committee recommended segregation and sterilization to preserve the blood of the American people.

The first eugenic sterilization law in the US was enacted in Indiana in 1907. It allowed for the sterilization of individuals deemed unfit to reproduce, such as criminals, the mentally disabled, and others confined to state institutions. Following Indiana's lead, other states began to pass similar laws. By 1927, 24 states had implemented eugenic sterilization laws, which targeted people with mental illnesses, disabilities, and those considered feeble-minded. The Supreme Court upheld the constitutionality of these laws in the case of Buck v. Bell in 1927, with Justice Oliver Wendell Holmes famously stating, "Three generations of imbeciles are enough." Eugenics became a rationale for sterilization laws, immigration restrictions, and other policies aimed at "improving the human stock," often targeting marginalized groups. By 1930, over 35,000 individuals were forcibly sterilized under eugenics laws. The number reached 60,000 in the 1970s.

Gradually, these ideas lost momentum. The last people to be condemned on the basis of anti-miscegenation laws in the United States were Richard and Mildred Loving. In 1958, Richard Loving, a white man, and Mildred Jeter, a woman of mixed African American and Native American ancestry, were married in Washington, D.C., to avoid Virginia's anti-miscegenation laws. Upon returning to Virginia, they were arrested and charged with violating the state's Racial Integrity Act of 1924. The Lovings were convicted in January 1959 and sentenced to one year in prison, with the sentence suspended on the condition that they leave Virginia and not return together for 25 years. They moved to Washington, D.C., but later sought to have their convictions and sentences vacated, arguing that the law was unconstitutional. Their case eventually reached the U.S. Supreme Court, which in 1967 unanimously ruled that anti-miscegenation laws were unconstitutional under the Equal Protection and Due Process Clauses of the Fourteenth Amendment. The Supreme Court's decision in Loving v. Virginia effectively ended all race-based legal restrictions on marriage in the United States.

About eugenics, in 1942, the Supreme Court ruled that the involuntary sterilization of certain inmates violated the equal protection clause of the 14th Amendment. However, this ruling did not affect those living in state-run mental health facilities. In the 1960s and 1970s, the civil rights movement and increased awareness of human rights led to growing opposition to eugenic sterilization laws. Public opinion shifted, and many states began to repeal or stop enforcing these laws. In 1974, Virginia repealed its Sterilization Act, and other states followed suit. In the 1990s and 2000s, several states formally apologized for their role in eugenic sterilization programs and provided compensation to victims.

Other countries besides the US got involved in social Darwinism and eugenic policies. Some provinces in Canada, such as Alberta and British Columbia, enacted eugenic sterilization laws, leading to the forced sterilization of thousands of individuals. In Europe, countries like Sweden, Norway, Denmark, and Finland implemented eugenic policies that included forced sterilization. In Sweden, for example, over 60,000 people were sterilized between 1935 and 1975. The UK had a strong eugenic movement, with influential figures and organizations promoting the idea of selective breeding to "improve" the population. While forced sterilization was not as widespread as in some other countries, there were significant efforts to discourage reproduction among those deemed unfit. Switzerland, too, implemented eugenic policies, including forced sterilization, targeting individuals with mental illnesses and disabilities.

The Soviet Union had some peculiarities in this area. The originators of Communism, Karl Marx and Frederick Engels, were enthusiastic followers of Darwin's ideas, and they saw parallels between the theory of evolution and their theory of historical materialism. Both saw class struggle as an embodiment of natural selection within society, and Marx saw Darwin's work as providing a naturalistic basis for understanding social change. However, both Marx and Engels wrote scathingly about "bourgeois Darwinians" who tried to justify social inequalities using Darwinian concepts. Engels argued that human society is shaped by social and economic factors, not just biological ones. In the Soviet Union, these ideas led to the prospect of creating a "New Soviet Man," partly based on eugenics but more focused on ideological and social conditioning rather than biological selection. But this idea never took hold in practice, and it doesn't appear that the Soviet Union ever applied eugenic policies in its history.

It was in the Fascist countries during the 20th century, Italy and Germany, that eugenics was most popular. Italy was more moderate. Italian eugenicists, influenced by Catholic teachings, focused on positive eugenics, which emphasized better prenatal care, education, and social welfare rather than coercive measures, so forced sterilization was never enacted in Italy. Nevertheless, the Italian fascists adopted racism as an official truth, and they acted according to the idea that Italians were no less Aryan than their German neighbors. Hence, the state forbade marriage with certain foreigners, including Jews, even though they were Italian citizens. Nevertheless, Italy never implemented the extermination measures that Germany carried out against the Jews and other ethnic groups. Still, the Italian government collaborated with the German government by sending Italian Jews to German extermination camps.

The Fascist attitude also implied that Italians had the right and the duty to dominate the dark races. The concept of a *Posto al Sole* (a place in the sun) in Africa was the exact equivalent of the German

Lebensraum (vital space) in Europe. This idea led to the invasion of Ethiopia in 1935, justified by the racial superiority of Italians over those dark people who happened to occupy a land that, for some unexplained reasons, really belonged to Italy. The Italian conquest of Ethiopia involved the extermination of a large number of Ethiopians, but, at least, their complete eradication didn't seem to have ever been planned, at least for the short period (five years) in which Ethiopia was a colony of Italy.

Nazi Germany, of course, was the country of eugenics. With World War I, the Germans were not only defeated on the battlefield but also starved by the Allies in a deliberately murderous food embargo that continued well after the end of the war. Shocked, impoverished, and humiliated, the Germans reacted the way human societies do when they are under heavy stress. They looked for a scapegoat, and they found it in the *Untermenschen*, the "inferior humans." They were supposed to be mainly non-German groups in Germany and the neighboring regions, but also German citizens who suffered from disabilities or who belonged to different ethnic groups, primarily the Jews.

The Nazi government implemented some of the most extreme eugenic policies in history, including the "Law for the Prevention of Genetically Diseased Offspring," enacted in 1933. It led to the forced sterilization of hundreds of thousands of people. The Nazi regime also used eugenic principles to justify the Holocaust, the systematic extermination of millions of Jews, Romani people, disabled individuals, and others deemed inferior and a threat to the state. Even ethnic Germans were targeted. Many of them were not only sterilized but physically eliminated because the very fact of being alive made them a burden to the state.

During the same period, Japan may not have been classified as a truly "Fascist" country, but its government had several characteristics in common with the Italian and German governments. In Japan, eugenics was connected with the diffuse perception of "racial purity." Nevertheless, the state intervened in the matter only after the end of World War II with the Eugenic Protection Law (EPL), in force from 1948 to 1996. It aimed to "prevent the birth of inferior descendants from the eugenic point of view and to protect the life and health of the mother." The law permitted the forced sterilization of individuals with intellectual or mental disabilities and those with certain hereditary diseases. Over 25,000 people were sterilized under this law, with around 16,500 of them not consenting to the procedure. In July 2024, Japan's Supreme Court ruled that the Eugenic Protection Law was unconstitutional and ordered the government to compensate the victims.

The misuse of Darwin's ideas was a tragedy of modern times, a result of the socio-political context in which his ideas were interpreted. In some quarters, it led to the demonization of the very concept of evolution by natural selection, and of Darwin himself, who never would have wanted to have anything to do with this distortion of his ideas. For the time being, fortunately, the ideas that were fashionable one century ago seem to have been abandoned, but you never know what the future can bring.

Contraception and Birth Control

In the Book of Genesis of the Bible, we can read the fascinating story of the young widow Tamar, who claimed her rights against all odds and succeeded in obtaining justice for herself and for her children. The story involves the actions of Onan, Tamar's second husband, who "spilled his seed on the ground" to avoid impregnating his wife. Today, we call that "Onanism", but that is more correctly described as *coitus interruptus*. It is one of the first mentions of contraception in history.

It would be long to explain why Onan did what he did, and why God then punished him with death. For what we are interested here, the story tells us that when Genesis was written, possibly as early as the 10th century BCE, people knew perfectly well that pregnancy was the result of full intercourse between a man and a woman, and that it could be avoided by refraining from performing or completing the act. Even earlier references to similar contraception methods can be read in some Egyptian papyri.

So, contraception is an old idea, but how old, exactly? Did it exist before written records? Were our hominin ancestors practicing it millions of years ago? We can be reasonably sure that most non-human creatures cannot link events spaced months apart and, probably, our remote ancestors couldn't do that, either. But, at some point in history, one of them must have experienced a "ha-ha!" moment: *that's* how children are born!

We can't know when that happened, but surely it was long ago. You can sometimes read that Australian Indigenous peoples did not connect pregnancy to sexual intercourse. It is a legend; a misunderstanding derived from some 19th-century reports. It is true, however, that some aboriginal groups interpreted human reproduction in terms of the combined effects of human sexual activity and spiritual or totemic forces. But that's nothing special. The concept that a God, or several Gods, intervene in causing the birth of a child is, after all, a common myth in many religions, including the Christian one.

Whenever the connection of sex to pregnancy may have been discovered, it opened up the possibility of conscious control of birth rates, and hence of population. Of course, the contraceptive methods available in ancient times were limited and sometimes dangerous. Abstinence was surely the most effective, but it was never popular. But a series of rules and social customs were often sufficient to keep boys and girls of reproductive age from having a chance to frolic together. One way was to forbid sex before marriage. A case in point is how Ireland managed to slash the country's fertility rate after the Great Famine of 1846 by developing social customs that encouraged late marriage.

In addition, there were several "soft" contraceptive tricks. Onan's coitus interruptus had the advantage that it required no special equipment. Another soft option was to refrain from sex only during a woman's ovulation period. This method is known today as the Ogino-Knaus method, from the names of its modern developers, Kyusaku Ogino (Japan) and Hermann Knaus (Austria). The problem with the Ogino-Knaus method is that human females do not provide signals to males that they are ovulating; a characteristic called "concealed ovulation." It is a reproductive trait that humans share with a few other species: dolphins, mice, and some birds. Bonobo females also partly conceal their ovulation. For the species that show this trait, sex is not just a feature of the reproductive mechanism, but a cultural and social behavior. As such, it can be used as a social communication tool. Bonobo females, for instance, use sex during their non-fertile period as a tool to control males. Human females can play the same game, but they can't easily detect their fertile period, even though they may have hints from various body signals that males cannot perceive. It means that the Ogino-Knaus method is not very reliable as a contraceptive method.

When soft methods failed, individual women and families had the option of abortion and infanticide. Both have a long history. In ancient times, abortion was very dangerous for the mother, and it seems that it was rarely practiced as a contraceptive method. Infanticide, instead, may have been more common. In some versions, it had an especially perverse ring, such as when, in Ancient Greece, weak or deformed newborns in Sparta were exposed on hillsides like Mount Taygetus. Another perverse form of infanticide was, and remains, the selective killing of baby girls. It is reported to be done to create a surplus of young males who have little chance of finding a mate except by killing the competition in war. In modern times, selective abortion or the killing of newborn females is reported to have become a problem in several regions of the world, when prenatal echography allows the early detection of the sex of a fetus.

A 2019 study[70] estimated 23 million "missing women" globally due to sex-selective practices, mainly abortion. Of these, China and India account for approximately 10 million each. In absolute terms, these are large numbers, but compared to the world's total population of 8 billion people, they represent less than a 0.3% difference. In evaluating these data, you also have to take into account that the natural ratio of females to males at birth is roughly 105 males to 100 females. It may be an evolutionary factor generated by the fact that males have a higher mortality rate than females. Not taking this factor into account leads to some exaggerated estimates, such as some newspaper articles that reported 63 million missing women in India, alone. [71] It is a number that simply makes no sense. Note also that the female/male ratio tends to decrease with time due to the longer lifespan of women. Hence, many countries have an excess of females in their population. For instance, Italy's population had a ratio of 95 males per 100 females in 2021. Nevertheless, Italy has a natural ratio of 105 males to 100 females at birth.

In extreme cases, society could react to a perceived unsustainable population excess by killing adults. We have evidence that killing the elderly was commonly practiced in the past and also in some contemporary societies. It even has a specific name among anthropologists: "gerontocide" or "senicide." The elderly were a favorite target for sacrificial rituals, and the usage may have been one of the factors that led to the legends about witches, often described as gaunt elderly women. Witches and witch hunts are characteristic of agricultural societies, and they still exist today and are practiced during famines. [72]

In our times, senicide is correctly seen as an abomination, but it may be closer to our world than we may imagine. During World War II, it was practiced in Germany when the government distributed cyanide pills to the elderly in order to press them to commit suicide to leave food available for those who could fight (I describe this practice in my book *Exterminations*, 2024). In some cases, societies may also target adults in their prime, and in this case, these killings have the aspect of ritual human sacrifices. Several wars, including modern ones, can be understood as "delayed infanticides," where large numbers of young men were directed to kill each other and cease to be a burden for society.

All these methods are unreliable, sometimes violent, and often traumatic. Modern contraceptive techniques are a much better option, and their development goes back centuries. Condoms are reported to have started being used during the 16th century, made with sheep intestines. In the 19th century, they started being made using natural rubber. In the 20th century, synthetic rubber was used, greatly reducing prices and making condoms one of the most common contraceptive devices in the world, with the added advantage that it provided a certain degree of protection against venereal diseases.

Intrauterine devices (IUDs) emerged in primitive forms also in the 19th century, and then they became much more effective and sophisticated during the 20th century. In 1960, the FDA approved the first oral contraceptive pill, Enovid, developed by Gregory Pincus, John Rock, and others. It combined estrogen and progestin to prevent ovulation, revolutionizing family planning. At present, many variants of contraceptive methods are available. Long-acting reversible contraceptives (LARCs), implants (e.g., Nexplanon), and improved IUDs have become popular for their convenience and efficacy. These methods provide reliable and effective ways to avoid undesired pregnancies.

Voluntary male and female sterilization appeared at the beginning of the 20th century but became relatively common in the mid-20th century. Data from the National Survey of Family Growth indicates that by the late 1970s, about 25-30% of contracepting women in the U.S. relied on female sterilization, while vasectomy accounted for roughly 10-15% of males' contraception. Globally, sterilization gained traction in countries like India and China, though often with government incentives that blurred the distance between voluntary and coerced.

Are contraceptive technologies part of the problem of the rapid fertility decline in some countries? It is possible, but, as always, it is difficult to analyze the behavior of complex systems when more than one factor acts to push the system in the same direction. Contraceptives make family planning easy, but, paraphrasing a well-known NRA slogan, "contraceptives don't prevent pregnancies, people prevent pregnancies." People who really don't want children can always find alternatives to rubber things, pills, and the like.

Nevertheless, contraceptives remain part of an ongoing cultural debate. Public sentiment toward contraception in 2025 varies globally. In some religions, contraceptives are seen as things that go against the will of God and hence are forbidden. Catholicism opposes contraception, but U.S. Catholic women often use it. Most Protestant groups support contraception, though some evangelicals debate its morality. Islamic scholars permit contraception within marriage, although there remain cultural stigmas in conservative areas (e.g., Pakistan). Hinduism and Buddhism have a generally permissive attitude. Traditional societies (e.g., rural Africa) may resist the concept of contraception due to beliefs in large families or spiritual concerns, or because they see it as disrupting fertility traditions.

In general, the question of contraception is strictly linked to the perception of population trends. When rulers needed a large number of young men to be used as cannon fodder (or pike fodder, or the like), contraception was abhorred and often forbidden by law. With the decline of the role of infantry in war, that ceased to be a priority, so that contraception became commonplace, not sanctioned, and even encouraged. With the current fertility decline, it may be possible that governments will backtrack and return to viewing contraception in a bad light, even forbidding it.

As always, the future is a labyrinth of bifurcating paths. We'll have to see where the many possibilities we have to control births at the individual level will lead us.

The Great Overpopulation Scare

Throughout history, overpopulation was a problem that ordinary people could perceive mainly from their empty stomachs. As a consequence, they would sometimes take the matter of reproduction into their own hands and quietly avoid providing rulers with the cannon fodder they required. That is, they would limit the size of their families by using whatever contraceptive methods were available.

For the rulers, instead, the idea that the commoners could decide by themselves whether to have children or not was anathema. The concept was considered both absurd and subversive. For them, there was a more straightforward way to solve the overpopulation problem: enlist the excess in the infantry and send them charging artillery and machine guns.

To see the concept of "overpopulation" appear as a societal problem, we have to wait for recent times. By the mid-20th century, data became available for the total world population. It was growing at a rate that seemed to be exponential. Many people saw the Malthusian tragedy in front of them: famines, wars, and new rounds of ethnic exterminations. It was a remarkable moment in human history; never before had the idea that there were too many people on Earth been a common worry and a subject of debate.

The overpopulation scare appeared as a typical dystopian theme of science fiction starting from the 1950s, reflecting the societal anxieties of the time. The stories tended to portray overpopulation through urban settings: endless city blocks and people packed into tiny living spaces. One of the first novels of this kind that reached the general public was Harry Harrison's *Make Room! Make Room!* (1966). It was later adapted into the film *Soylent Green* (1973), which depicted a dystopian 1999 New York City with 35 million inhabitants living in squalid conditions. There followed novels such as William F. Nolan and George Clayton Johnson's *Logan's Run* (1967), which explored the theme of a society that obtained population control by eliminating people as soon as they reached 21 years of age. In 1968, John Brunner published his novel *Stand on Zanzibar*, with the title being based on the idea that the whole population of the time could fit standing shoulder-to-shoulder on the island of Zanzibar. Perhaps the most impressive story of this period was Robert Silverberg's *The World Inside* (1972), which depicted a dystopian world where people had given up all their rights except one: that of having sex and reproducing at will. Billions and billions of people lived in densely populated, gigantic skyscrapers, where all dissent was punished with immediate death, but where sex was completely free for everyone as desired.

Overpopulation fiction was merely a reflection of a theme that was also explored in scientific terms by individual researchers and research institutions. In the 1950s, the Rockefeller Foundation funded research on population growth and its implications. John D. Rockefeller III was a prominent advocate for population control. In 1952, he stated: "Population growth is the most serious problem facing mankind today." In the 1960s, the United Nations began to address population growth as a global issue. A 1963 UN report[73] stated: "The rapid growth of the world population is one of the most serious problems facing humanity, with profound implications for economic development, resource use, and environmental sustainability."

Among individual researchers, William Vogt's 1948 book *Road to Survival* was one of the first works to explore the modern concept of population growth and the probable outstripping of the available resources. He wrote that "we are already far beyond the Earth's carrying capacity." And "The day of reckoning is close at hand. The human race is caught in a dilemma of its own making." His ideas remained influential in the 1950s and 1960s as fears of resource depletion grew.

Fairfield Osborn's 1953 book *The Limits of the Earth* argued that population growth would lead to environmental degradation and resource scarcity. In 1954, Hugh Moore published a pamphlet titled *The Population Bomb*, which later inspired Paul Ehrlich's more famous 1968 book with the same title. Ehrlich predicted widespread famines in the 1970s and 1980s. He wrote, "The battle to feed all of humanity is over... hundreds of millions of people will starve to death." Isaac Asimov, the well-known science fiction and science popularization author, wrote several essays on overpopulation. In 1988, in an interview with Bill Moyers, he said, "Democracy cannot survive overpopulation. Human dignity cannot survive it. Convenience and decency cannot survive it. As you put more and more people into the world, the value of life not only declines, but it disappears." Eventually, all the worries of this period could be seen as a revival of Malthus' ideas, and the proponents were sometimes described as "neo-Malthusians." But this modern version was based on a concept that Malthus didn't consider and that, probably, he couldn't even conceive: overexploitation, or, with the same meaning, "overshoot."

The idea of overexploitation is a cornerstone of biological population models. It implies that a population grows proportionally to its current size and, at the same time, to the amount of available resources. At some point, the population size may exceed the capability of the resource stock to replenish itself. At this point, the stock starts dwindling; nevertheless, the population growth can continue for a while, entering the condition called "overshoot." Eventually, the resource stock becomes unable to sustain the growing population, and the result is collapse.

The concept was implicitly expressed in the biological models independently developed in the 1920s by Alfred Lotka and Vito Volterra, sometimes defined as the "Rabbits and Foxes" model.

Imagine an island inhabited only by foxes and rabbits. The fox population tends to grow in proportion to their food source, the rabbit population. Unfortunately for the foxes (and also for the rabbits), the number of foxes tends to become large. The result is that they will kill off nearly all the rabbits. At that point, the foxes will be in full overshoot. Having destroyed their food source, they will necessarily starve and die. The Lotka-Volterra model assumes that with the foxes nearly gone, the few surviving rabbits will be free to reproduce and the cycle will restart. In actual ecosystems, a population crash may lead to extinction. Human beings are not immune to this destiny, as discussed in William Catton's book, *Overshoot* (1982)

Probably, the most influential and most discussed population study in this period was *The Limits to Growth*, published in 1972 and created by a group of MIT scientists. It was the first to attempt a quantitative evaluation of the world's economic system, taking into account an ensemble of parameters that included population. The study sketched various possible methods to stave off the economic collapse envisioned for the first decades of the 21st century, keeping the system at a moderate level of activity that was presumed to be possible to maintain for at least the whole 21st century and probably much longer, while avoiding a population collapse. You can find more details on the study in the appendix of this book.

Of these proposals, those that involved putting a cap on economic growth and consumption were rejected almost immediately by politicians and economists. The standard view in economics was (and remains today) that economic growth is always good and always to be pursued. Doing otherwise looked heretical, strange, and unspeakable. No politician could ever be elected by promising people to reduce their consumption, and hence their wealth. The idea fell flat, even though a small group of heretics supported (and still support) consciously choosing "degrowth" as a way to manage the economy.

Conversely, the idea of acting on the population by birth control looked easier and more palatable. It seemed logical that fewer people would mean more resources per person, and that would make voters happy. At the same time, it would reduce the strain on depleting resources. Hence, economic growth could be maintained, at least in terms of GDP per person. Unfortunately, this idea didn't agree with the results of the Limits to Growth study. The model could be used to calculate the future of the economy for a situation in which the global population was kept constant at the 1972 levels. The result was not encouraging. If population control was coupled with policies aimed at reducing consumption, then the economic system could be stabilized. But if population was the only parameter that was kept stable, the model showed that collapse would occur anyway. True, the consumption per capita would increase to much higher levels than in the run without population control. Still, the strain on resources and the pollution problem would scarcely change, and collapse would occur anyway. The difference was that people would be much richer at the start of the collapse, but just as poor once they reached the bottom.

Nevertheless, the idea of solving complex problems with simple solutions is always attractive to decision-makers and the public alike. So, starting in the 1950s, the idea that population control was a magic pill to cure civilization's ills started being discussed and proposed by several pressure groups and associations. In most cases, the concept of "population control" was carefully framed in terms of "birth control," supposed to be painless, non-violent, and a choice, not an obligation. Considerable efforts were spent to distance these efforts from the earlier abominable methods used by the German Nazis. In time, the term "birth control" was almost universally replaced with the less threatening one of "family planning."

Initially the campaign for birth control/family planning affected countries that had high population density and a recent history of famines. In 1952, India became the first country to launch a national family planning program. The government promoted various contraceptive methods to reduce fertility rates, including male sterilization. The program was resisted in rural areas where men saw it as a threat to their masculinity, while it was more successful in urbanized regions. In any case, the average total fertility rate in India declined from nearly six children per woman in the 1950s to the current value of about 1.9, below the population replacement rate. This result may not be due to the government's efforts, but rather to the population transition that appears to occur everywhere, regardless of what the government does or doesn't do.

Other countries faced similar problems. China was the one that implemented the most drastic measures to reduce natality by forbidding families from having more than one child. This story is discussed in a special section on China in this book. Other countries implemented family planning initiatives based on contraceptives, sterilization, and the legalization of abortion. The list would be long, but some notable cases are Indonesia and Bangladesh, which adopted measures similar to those adopted in India. In Africa, memories of the European colonial invasions made governments wary of acting to reduce population, an idea they saw as an invitation to be invaded again. Nevertheless, efforts to control population growth are being made in Kenya, Ethiopia, Rwanda, and Senegal. Still, the results have not been very successful, and Africa remains one of the few regions of the world where the human population continues to grow rapidly.

In Latin America, the efforts have often involved contraceptives, and some regions have an exceptionally high rate of use. For example, countries like Colombia, Peru, and Paraguay have achieved high levels of modern contraceptive use, ranging between 51% and 70%. In Latin America, the Mexican President, Luis Echeverría (1970-1976), implemented some policies to reduce the number of children, at the time an average of 8/10 per woman. Apparently, this was done as a direct result of the work by Victor Urquidi, an economist, academic, and member of the Club of Rome. In this case, and likely only in this case, the Club of Rome directly influenced a major country's population policies. In Europe, not many countries adopted specific family planning actions, and the idea was abandoned as soon as they realized that they were facing a serious problem of fertility decline.

In the US, before World War II, the state encouraged people to have large families. Contraceptives were prohibited by law at various points in history. For example, the Comstock Act of 1873 banned the distribution of "obscene" materials through the mail, including contraceptives. It wasn't until the landmark Supreme Court case, Griswold v. Connecticut, in 1965, that the ban on contraceptives was overturned, at least for married couples. This decision was extended to unmarried individuals in the 1972 case Eisenstadt v. Baird, which established that the right to privacy includes the right of individuals to use contraception. The use of the contraceptive pill was approved in 1960, and in 1965, the Supreme Court declared unconstitutional state laws prohibiting contraceptive use by married couples. In 1973, the Supreme Court case Roe v. Wade legalized abortion.

But it was a brief moment. Birth rates rapidly declined in all areas that were called the "First World" (the West) and the "Second World" (the Soviet bloc and China). This trend was known as the "demographic transition" and didn't require draconian government measures to occur. As a result, family planning was rapidly losing momentum, while a strong new push for natalism appeared. The last hurrah of the anti-natalists may have been the 1992 declaration of the "Union of Concerned Scientists."

> "We, the undersigned, senior members of the world's scientific community, hereby warn all humanity of what lies ahead. A significant change in our stewardship of the Earth and the life on it is required, if vast human misery is to be avoided and our global home on this planet is not to be irretrievably mutilated."[74]

The declaration is still online, but the changing perception of the issue led the UCS to precede this document with a recent preface that defines it as "deeply problematic."

The Return of Natalism

At present, the perception of overpopulation as a problem remains diffuse. It is a correct stance considering that humankind has passed the 8 billion mark and is still growing. On the other hand, this perception has never been shared by everybody. The modern clash between the natalist and the anti-natalist positions became clear with the World Population Conference organized by the United Nations in Bucharest in 1974. It was supposed to be a milestone in which contraceptives and population control were to be diffused and promoted everywhere in the world. Instead, the representatives of Third World countries completely rejected the idea.

At that time, the concept of "Population Transition," that is an impending decline of the human population, was not yet evident from the available data. But a fertility decline had already been noted in many countries. So, a key slogan that emerged from the conference was "Development is the best contraceptive." This position argued that the key to addressing population growth was not through coercive family planning or birth control programs, but rather through socio-economic development. The proposers believed that by improving living standards, education, healthcare, and women's status, population growth would guarantee maximum prosperity for everybody.

The position of the Third World nations soon had the upper hand in the debate, with the proponents of population control accused of planning dark and dire things such as exterminations and genocides. Still today, the term "population control" has a ring of evil attached to it that seems to be impossible to remove. Note, however, that a few years after the conference, China took a completely different position, favorable to birth control. The case of China will be discussed in a specific chapter in this book.

The debate went on, and the clash of natalism and anti-natalism continues to this day. It is mainly a political debate, and, as with all political debates, it follows the rules of politics, the main one being that the way to win is to demonize your adversary. One of the first targets of demonization was Thomas Malthus and his *Essay on Population*. Malthus had already been heavily criticized in his time, accused of undermining faith in divine providence, arguing that his ideas painted God as a harsh taskmaster rather than a benevolent creator. But the modern campaign turned the term "Malthusian" into an insult. Just as an example, Matthew Simmons, an expert in petroleum economics, reported that he gave a talk in 1994 in which he mentioned *The Limits to Growth* as not being obviously wrong, and that afterward, one of the attendees told him that he was "obviously a Malthusian" and, for that reason, wrong[75].

If you have some experience with giving public talks about these subjects, you know how easily it is to hear a simple (and wrong) syllogism that goes as, "Since what you say sounds like what Malthus said, and since Malthus was wrong, then you are wrong, too." We might refer to this approach as "Hitler's vegetarianism fallacy." That is, "Since Hitler was vegetarian, and since Hitler was evil, then vegetarianism is evil." But this is the way things go in the political debate. Anything you say can be twisted out of context and turned into something that condemns you to shame among men.

Here is a typical example of how Malthus was attacked using rhetorical tricks and distortions. Joel Mokyr, in his book titled *Why Ireland Starved* (1983), reports this sentence from a letter that Malthus wrote to his friend David Ricardo,

> The land in Ireland is infinitely more peopled than in England; and to give full effect to the natural resources of the country, a great part of the population should be swept from the soil.

This sentence gives the impression that Malthus was advocating the extermination of the Irish. But the actual sentence that Malthus wrote reads:

> The land in Ireland is infinitely more peopled than in England; and to give full effect to the natural resources of the country, a great part of the population should be swept from the soil into large manufacturing and commercial Towns.

Clearly, Malthus wasn't proposing to kill anyone; rather, he proposed the industrialization of Ireland to create prosperity in the country. Nevertheless, legends spread quickly on the web. You can see the truncated sentence by Malthus repeated over and over in other papers and on the Web to demonstrate that Malthus was an evil person who proposed the extermination of the poor. You can even read that this sentence demonstrates that the Irish famine was Malthus' personal fault. Too bad that at the time of the famine, Malthus had been dead for more than 10 years.

A similar treatment was reserved for the 1972 study *The Limits to Growth,* which was accused of being ideologically bent toward exterminating the dark races and having made "wrong predictions." You can read the details in my 2011 book *The Limits to Growth Revisited*. In time, the thesis and the data of The Limits to Growth seemed to fade from people's minds. However, there remained plenty of legends accusing various persons and groups of plotting a global culling of humankind or some specific ethnic group. The ancestor of these theories is *The Protocols of the Elders of Zion,* a forged document which surfaced in Russia in 1903. It described a secret Jewish cabal plotting world domination, including the destruction of the Christian civilization, explicitly framed as an extermination. But these early legends pale in comparison with the hyperactive theorists engaged in finding extermination plots everywhere.

Despite multiple debunkings, the idea that culling plans are being prepared against humankind by evil secret societies remains very much alive. As culprits, various existing organizations are commonly listed: the World Economic Forum, the Club of Rome, the United Nations, HAARP, the Federal Reserve, the Bilderberg Group, and more. Others are creations of pure fantasy: The Gnomes of Zurich, the Reptilians, the Illuminati, and more. Even some individuals are often accused of covertly planning to cull most of humankind. The most often targeted one is probably Bill Gates, but the chairman of WEF, Klaus Schwab, is probably well placed in the second position. Currently, Elon Musk is gaining popularity as a would-be humankind exterminator. Yet, he is also known for his many statements about the need for women to have more children.

This binary attitude is diffuse throughout the debate and is likely to become a major stumbling block in all attempts to manage a demographic situation, which, as with everything that has to do with Earth's ecosystem, would need a nuanced approach, avoiding extremes such as terrorizing people because of overpopulation, or, for the reverse effect, extinction.

At present, the "natalist" position seems to have gained the upper hand, and most governments are making efforts to increase birth rates by supporting families with financial and other kinds of help. These efforts, so far, are failing, but perhaps more effective methods will be developed. Yet, all that's done in this field is based on the idea that low birth rates are the result of sociological factors only. The possibility that they are instead caused by pollution is almost never mentioned. That's probably because, if it turned out to be the main factor, or even just an important one, very little could be done to restore women's fertility, at least in the short run.

In the end, our future will depend on how we can avoid radical reactions to a situation that is rapidly evolving. There are plenty of mistakes that can be made in the attempt to control birth rates, and it doesn't matter much in which direction. Mistakes have been made in the past, and will probably be made again in the future. But this is how the human mind works, and whatever happens to us in the future will be, in large part, a result of our actions.

6. Case Studies

The faces of these people were subdued with hunger; pale, or rather of a ghostly yellow, indicative of the utmost destitution. They are starving. We hurried with horror from these frightful visitations, which are permitted by Providence for his own wise ends, sick at heart. (Marquis of Waterford, April 1846)

Figure 15 - Bridget O'Donnel and her two children at the time of Ireland's great famine. Wikipedia, Public Domain.

This section describes some historical cases that showed interesting population trends, either in terms of rapid decline, stability, or growth. It is not supposed to be a complete listing of all known cases. Still, it provides several examples that may be useful as an illustration of the considerations developed in the previous sections of the book.

Ancient Rome: metal poisoning

The ancient Romans had many things in common with us, but they never had a problem of overpopulation. Just consider that at the height of the Empire, ca. 1st-2nd century CE, the population of Western Europe is estimated to have been 20–30 million. Today, it is more than 10 times larger, with over 300 million people. From our viewpoint, Europe would have looked empty: a land of forests and small towns, crisscrossed by the network of Roman roads which had more of a military purpose than a commercial one. During Imperial times, overcrowding problems were mostly unknown, even though the capital city, Rome, had become large enough that it had traffic jams. Yet, the agricultural production of the time was more than sufficient to provide food for Rome and for all the cities of the Empire. It was the result of a sophisticated maritime transportation system that delivered food from the fertile regions of North Africa and the Middle East all the way to the coastal cities. The system remained in function up to the mid-5th century CE, as long as the Empire existed, securing food for the Roman citizens.

If anything, the ancient Romans had a problem of underpopulation, and they understood that perfectly well. With the start of the economic decline of the Empire, during the 1st-2nd century CE, natality plummeted, and it became more and more difficult to find enough manpower for the legions and the state apparatus. Several emperors enacted laws that encouraged couples to have children through financial incentives and tax breaks. Augustus (1st century BCE) was probably the first ruler to develop the concept of "giving sons to the country," a concept amply exploited by later rulers who saw themselves as extreme cases of alpha males, "fathers of the country." However, despite the emperors' efforts and the attempt to raid as many slaves as possible from neighboring countries, the Roman population started declining in Imperial times.

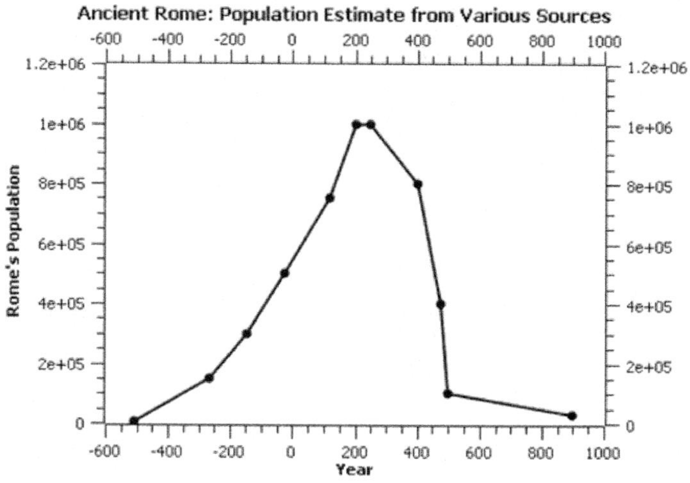

Figure 16 -- An approximate reconstruction of the population of ancient Rome. Data from various sources

In the figure, you can see a graph of the population of Rome. Take it as no more than an illustration: the data on the Roman population are highly uncertain and intensely debated among specialists. Nevertheless, the results are consistent with other reconstructions. The Roman population grew together with the extension, wealth, and power of the Empire. Then, it went down rapidly. It was a clear example of a trend that the Roman philosopher Lucius Seneca had identified during the peak of the Empire's power: that fortune grows slowly, but ruin is rapid. I called it the "Seneca Effect."

Eventually, lack of manpower made it impossible for the Romans to defend and maintain their provinces, and they had to abandon most of them. Finally, they were invaded by Germanic populations, who were not necessarily hostile but replaced the vacuum left by the Roman demographic decline. During the 5th century CE, the Western Empire faded away forever. Afterward, the European population is said to have been reduced to as few as 18 million people.

The collapse of the Roman Empire has puzzled historians for a long time, and hundreds of "explanations" have been proposed for an event that looked improbable, or even impossible. The mighty Roman legions were powerful and nearly invincible for almost a millennium. Then, in the 3rd century of our era, they started losing one battle after another, and in a couple of centuries, they had disappeared. The old Empire left only ruins and mentions in history books.

Most of the proposed reasons for the collapse were based on qualitative arguments and the whim of the historians who proposed them. "Loss of moral fiber," "influence of oriental religions," or "immigration," and many more. Only some proposals were based on real-world data. Of course, we lack detailed statistics about how the Roman economy evolved, but we can use proxy data. For instance, we can gather information on the productivity of the mining of precious metals from the fraction of silver in the Roman denarius coin. These data indicate that silver production collapsed with the great crisis of the 3rd century. A reasonable explanation is that the exhaustion of the Roman mines reduced the supply of gold and silver to the Imperial coffers. That made the Empire weaker and cracked the very fabric of the Roman society, which was based on monetary exchanges, as well as on the capability of paying fighters with hard cash. Not for nothing, the modern term "soldier" derives from the ancient Roman term "Solidus" for a gold coin of the late Empire.

The demise of complex systems, such as human societies, is rarely generated by a single factor. So, the depletion of the Roman precious metal mines surely played a role, but it may not have been the only cause of the collapse. An interesting peculiarity of the Roman collapse is that it may be the only case in history of a pollution-generated population decline for which we have quantitative data. We know that the Ancient Romans suffered from poisoning by heavy metals, in particular lead and mercury. The idea that the Romans weakened themselves because of lead poisoning was a remarkable intuition by a German scholar named Rudolf Kobert in 1909. He lacked the tools to prove his ideas, and his theory was mostly ignored. Later, it was reproposed by S. C. Gilfillan in 1965.[76] Today, historians still debate this matter, but there is no doubt that heavy metal poisoning was a feature of the late Roman Empire. The Roman economy was one of the most successful mining-based economies in ancient history. They mined all sorts of metals, and their wealth was based on their production of gold and silver. But they also mined non-precious metals, including poisonous ones: lead and mercury.

Lead was widely used in Roman times for many practical purposes. Easy to smelt and workable at room temperature, lead could be used for pipes and vessels of all kinds. In Imperial times, drinking wine from lead cups became fashionable. The slight corrosion of the lead by the wine generated dissolved lead salts, giving the wine an extra sweet flavor that the Romans appreciated, but it also poisoned them. They may have understood that lead had harmful effects on human health, but they seem to have associated it only with the people who smelted lead. These workers were often affected by a mental degradation called "Saturnism," from the association of lead with the Saturn planet and the divinity Saturn. It doesn't appear that the Romans ever understood that their fancy wine goblets were making them sick. And that was not the only problem they had with lead. It was extensively used for pipes, vessels, and kitchen tools, and that generated a further source of lead that poisoned the Romans. It was also released into the air when used to smelt silver and gold from the rocks in which they were stored.

We know about lead poisoning in Roman times because lead is a systemic poison that accumulates in the bones. Its chemical properties are similar to those of calcium, a normal component of bones (it is termed a "calcium analog"). Nowadays, we can measure the lead content in bones by examining ancient Roman skeletons, and in this way, we can have some idea of how bad the problem was. [77] The result is that the Romans had enough lead in their bones to cause serious problems such as anemia, high blood pressure, kidney disease, and fertility problems. High levels of lead are also associated with brain damage. Considering these effects, we may conclude that lead poisoning had a significant effect on the health of the Romans, and some historians speculated that the erratic behavior among Roman elites and emperors might have been due to lead poisoning. Did Caligula make his horse a senator because he drank too much wine in lead goblets? We will never know for sure, but we cannot exclude it.

Mercury was another poisonous metal that the Romans extracted, although not in large quantities. They used it in pigments, in particular as a cosmetic powder in the form of cinnabar (mercury sulfide). They also used it for gold extraction using the "amalgam" process, and for ceremonial purposes. Mercury is a fascinating substance: the only metal that is liquid at room temperature, but it is also a serious systemic poison for the human body. The Romans were aware, at least in part, of the toxic effects of mercury. The warders of mercury mines knew that the slaves engaged in mining would likely succumb to a mysterious sickness. The symptoms were tremors, extreme mood changes, and eventually loss of hearing and restricted vision, and the people affected were defined as having a "mercurial" behavior. Archaeologists have found significant traces of mercury poisoning in ancient Roman skeletons.[78] Again, mercury may have been an important factor in sickening them and causing the demise of the Empire.

The similarities with our society are, at the same time, numerous and worrisome. We are producing and spreading over the ecosystem a multitude of substances whose effects on the human body are scarcely known or even totally unknown. How about rare metals such as platinum, palladium, rhodium, and several others? They are so rare that humans have never been in contact with them. And, right now, we are spreading them in the environment as a result of the degradation of catalytic converters used to abate other poisonous substances in car exhaust. There are many more substances released in the ecosystem whose effects are scarcely known, just as their combined effects. Yet, it doesn't seem that the authorities or the public are overly worried about the problem. Just like the ancient Romans, we are poisoning ourselves without even realizing what we are doing.

Could some astute Roman philosophers have noted that the Romans were killing their own population by their smelting activities? Could they have tried to alert the authorities about the problem that was risking the destruction of their world? If someone did, we may imagine that they were the target of a demonization campaign by the lead and mercury smelting lobby that succeeded in presenting them as slightly feebleminded catastrophists, whom nobody should listen to. Of course, that means using some fantasy to fill the missing historical records. Still, it is also true that the Romans had developed methods for "perception management" (aka "propaganda"), which may not be as refined as our modern ones but were nevertheless effective. And this is another worrisome element of similarity with our modern world. Are we going to be killed by propaganda telling us that some things are harmless, while we should know they are not?

Medieval Europe: Plagues and Famines

The COVID-19 pandemic that struck the world in 2020 was an exceptional event in human history. Never before had the entire world reacted to an epidemic so strongly and with such extreme measures. Never before had people accepted being locked inside their homes, forced to wear face masks all the time, to avoid getting close to other people, to be tracked, examined, tested, and provided with "passes" to be able to perform their normal activities. But many of them were genuinely terrified by a virus that, at the beginning, was described as an existential threat to humankind.

The wave of terror that swept the world in 2020 century may have been a reflection of the great scare that originated from the disasters that struck Europe during the 14th century. First, there came the Great Famine of 1315–1317, then, 30 years later, the epidemic termed "Black Death." Both took their toll on the European population, estimated to have fallen from roughly 75 million in 1300 to 25–35 million by 1400, representing a 60–70% total decline over the course of the 14th century. These numbers are uncertain and may be exaggerated, but the disaster had truly apocalyptic proportions.

Figure 17 - Totentanz," the dance of death (1493) by Michael Wolgemut. Public Domain

The famine not only impacted the European population, but also the way of thinking of Europeans. It was at that time that painters started producing images of the Totentanz, the dance of death, with animated skeletons shown as armed with swords and pikes, overwhelming the puny efforts of living humans to stop them. It was at that time that it became fashionable to have tombs where the image of the buried person appeared as a cadaver. For a few centuries, European thought seemed to be focused on death and the decay of the body. It was the time when the concept of *memento mori* (Latin words for "remember that you must die") became commonplace.

What happened that caused this huge disaster? We tend to see the Middle Ages in Europe as a "dark age" of superstitions and barbarism, but reality was very different. After the collapse of the Roman Empire in the 5th century CE, Europe had rebounded to create new social structures, a new economy, and a new culture based on Christianity. With the discovery of new silver mines in Eastern Europe, the economy could be "re-monetized" after the collapse of the Roman currency, and that generated a new phase of rapid economic growth. The population was growing, too, and, with the turn of the first millennium, it was clear that Europe had an excess of young men willing to fight for money and glory. That was the origin of the Crusades, when Europeans found no better way to get rid of the marauding bands of knights than sending them to fight in the Levant. That phase of history began in 1095 with the first military expedition attacking the Holy Land, marked by the cry, *Deus Vult*, meaning "God wills it."

The age of the Crusades lasted about two and a half centuries. It ended in 1291, when the last European stronghold in the Middle East, the city of Acre, was abandoned. It was an unmitigated disaster for Europe. Its economy had been wrecked by the war effort, and its forests had been destroyed to make ships and weapons. Now, the exhausted soil couldn't produce sufficient food. The Great Famine struck in 1315.

The great famine is not usually attributed to overexploitation but, rather, to climate effects. The chroniclers of the time noted how, from 1315 to 1317, heavy rains flooded fields, rotting seeds before they could sprout. The pattern was repeated, although less severely, in the subsequent year, and the famine didn't completely subside until 1324.

This series of events is well known, but is it true that the famine was caused by a change in weather patterns, as it is normally interpreted?

Of course, in those times, rainfall measurements were not performed, so the chroniclers could only discuss local effects and qualitative narrations. Today, we can do much better, and the data obtained from dendrochronology (the study of tree rings) does not show significant changes in rainfall before or during the famine.[79] It is more likely that the effects attributed by the chroniclers to higher-than-normal rainfall effects were instead due to soil degradation, in turn, caused by deforestation. Indeed, deforestation is clearly detectable in the dendrochronological data.

Apparently, in the 14th century, Europe was in a condition not unlike the current one, where the effects of normal rainfall are enhanced by the impermeabilization of the soil caused by deforestation and buildings.[80] The results may be disastrous floods and landslides, today as in ancient times. In a certain sense, the Europeans themselves had created the disaster that struck them in the form of the Great Famine.

The Famine was bad enough, but it was the origin of an even worse catastrophe: the later epidemic called Black Death. It is known that when a population is weakened by a lack of food, it is easy prey for all kinds of pathogens. Among many examples in history, this happened after World War I, when most of Continental Europe had been suffering under the effects of the Allied sanctions, preventing the import of food from overseas. The weakened population was struck by the Spanish Flu, which killed millions, starting in 1919.

In the case of the Medieval European Famine, malnutrition persisted for several years after the date chosen by historians to mark its end. Skeletal evidence reveals that people born during and shortly after the famine suffered from stunted stature and compromised health.[81] Hence, they may have been ill-equipped to resist the pestilence that hit them. The changes in the landscape, too, may have facilitated the spread of the epidemic. Depleted herds, fallow fields, and disrupted trade slowed recovery and forced peasants to flood into towns, which became overcrowded and hence a probable focal point for the spreading of pathogens. This correlation may explain why the pestilence hit so hard, despite having been endemic in Eurasia for at least one millennium before.

So, it happened that, in 1347, a ship coming from the Middle East docked at Messina, Sicily, its crew afflicted with fever and grotesque swellings. It was the start of the epidemic that was later called the Black Death. The infected people persons suffered fevers, chills, and painful buboes. These symptoms were followed by hemorrhagic coughing or the blackened decay of gangrene. Few survived beyond days. Within four years, by 1351, the plague had swept all over Europe, claiming an estimated 25 to 50 million lives. Entire settlements vanished; large cities saw the loss of a large fraction of their inhabitants, so many that unburied corpses littered the streets. The chroniclers report of deserted villages, fields overtaken by weeds, and wolves roaming where herds once stood.

To contemporaries, the plague was no mere illness but a retribution for people's sins coming directly from the Lord. Clerics like John of Reading deemed it the prophesied end of the world. Flagellants scourged themselves in penance; others accused Jewish communities, inciting massacres across Europe, amplifying a trend of pogroms that had already started during the First Crusade. Physicians speculated on miasmatic vapors or planetary conjunctions, advocating bleeding as a cure, often hastening death.

Attributing the Black Death to God's wrath was no more accurate on the part of the chroniclers of the time than attributing the Great Famine to excessive rain. Modern science identified the culprit of the plague: a bacterium, the Yersinia Pestis, whose origins were likely in Asia's steppes. Its name comes from Alexandre Yersin, a Swiss/French physician who discovered it in 1894. The bacterium can affect fleas, which attack rats or other rodents. From the rat, it can spread to humans through the bite of infected fleas. When that was discovered, rats gained a bad reputation that persists to this day. But, in practice, the bacterium can spread to humans much better in its airborne form as respiratory droplets than by the rare event of a rat biting a human being. Besides, rats can't travel as fast and far as human beings do. God, and even rats, had little to do with the Black Death, but the disaster surely had biblical proportions.

The Great Famine and the Black Plague are examples of how a population can collapse continent-wide. It was the result of several coupled factors, but one of them was overpopulation. Of course, Europe in the 14th century had a much smaller population than it has now. However, overpopulation is a term that depends on circumstances. For the capability that people had to manage their territory at that time, 80 million people were too many. It was not a Malthusian collapse directly caused by the limits of food production. We may see it more appropriately as a pollution-related collapse, with the term "pollution" intended to refer to the degradation of the ecosystem caused by human activities, such as deforestation.

It is sobering to think that we now have about 750 million inhabitants in Europe, intended as the geographic region that reaches all the way to the Urals, almost ten times more than in the 14th century. We have much better technologies that are supposed to help us cope with epidemics, but would it be possible to see a new outbreak of a major plague? It looks unlikely, at least in the current situation. Consider that the Black Death killed 20–50% of Europe's population. Compare to the COVID-19 toll: 0.09% of the global population. The fame of a global scourge of the modern pandemic was much exaggerated.

Today, we seem to be able to resist pathogens much better than our ancestors because of a combination of better nutrition, better hygiene, better medical care, and vaccines (even though the last category has fallen in disrepute with many people nowadays). In addition, the habit of long-range traveling in modern times has exposed people to contact with populations that were historically separated. It means that our immune system is continuously exposed and tuned to new microorganisms coming from far away. That gives us time to develop immunity.

So, maybe we don't have to worry too much about a new black plague, at least as long as we can have a good standard of living, allowing us to remain healthy. Even the famed "bioweapons," object of so many worries at the time of the invasion of Iraq in 2003, may have been an overhyped threat. So far, the only case of a biological weapon having caused victims is that of the spreading of anthrax pathogens in the US in 2001, after the 9/11 attack on the World Trade Center. To this date, we don't know who was behind the attack, but we know that it resulted in a grand total of five victims. Consider also that the COVID-19 virus was said to be a bioweapon escaped from a research lab. If it was so, it was no better than a high-tech rubber band launcher. Things may change, though, in the hypothesis of a societal and economic collapse, which would leave people defenseless against normally not worrisome germs. It would be part of the "feedback" concentration effect that dynamic models see as a possible cause of a rapid population decline. The future is, as always, uncertain.

Ancient Europe: Inheritance and Population

Fertility and wealth are two strictly related factors in human populations. Wealth is accumulated over time, but it is well known that you can't take it with you when you start your path toward the Elysian Fields. So, if wealth is divided equally among many heirs, it would be gradually fragmented and dispersed, especially in times of high natality. One measure to counter this dispersion was the idea of "primogeniture," that is, the concept that a single heir, typically male, would take over the whole of the family's wealth. This use was diffused all over the world, but it was especially important in Ancient Europe, from Medieval times to relatively recent ones.

Primogeniture was a nearly perfect "zero-sum" game: the family's wealth remained within the family. Playing this game, however, was difficult for various reasons; the main one being that the disinherited sons and daughters of a rich family were not happy about the idea, understandably so. That could have been solved by having just one male heir, but that was impossible in times when contraceptive techniques were ineffective and dangerous, and child mortality was high. A wealthy family aiming at just one heir could see him die before having a chance to marry and have children. To avoid that, families needed to have more than a single male child, but that created a serious tension inside families. You may remember Alexandre Dumas' *The Man in the Iron Mask* (1847), a novel centered on a mysterious prisoner held in the Bastille during the reign of Louis XIV. He turned out to be the twin brother of King Louis XIV, hidden away to secure Louis' throne. An extreme case, but a dramatic illustration of the inner conflicts that could arise among brothers inside wealthy families.

About disinherited daughters, you surely remember the story of Cinderella which exists in various versions all over the world. All of them deal with the fate of a girl whose parents couldn't or wouldn't provide her with a dowry, preferring to concentrate their resources on another daughter. These stories often end with the dispossessed girl finding a trick to marry her Prince Charming. However, in the real world, it wasn't so easy to find a good fairy ready to turn a pumpkin into a coach. The problem existed in all pre-modern cultures, and we find it mentioned already in Sumerian texts that go back five thousand years ago.

Disinherited sons could still build up a career in an expanding economy by becoming warriors, explorers, or merchants. The dispossessed daughters, too, had ways to react to their bad luck. They could still find a comfortable position in society if they could marry a promising hotshot, even though not necessarily a rich one. Things were different in a static or contracting economy, such as in early Medieval Europe. In these conditions, the penniless scion of a wealthy family could become a knight errant and engage in the search for enemies, possibly in the form of dragons. That gave rise mainly to legends and poems, but in the real world, dragons were rare, and unemployed warriors were a nuisance for everybody. They were often gotten rid of by sending them on some military adventure where they could unleash their homicidal instincts against foreigners. It is still a common strategy nowadays.

About disinherited girls, they wouldn't usually become unruly warrioress themselves, despite the many modern movies depicting sexy female fighters. But they were nevertheless a disruptive element of society, scheming to seduce powerful leaders. The resulting children were termed "bastards," not necessarily an insult in ancient times, but an element of unrest as potential claimants to the leader's wealth.

One way to avoid the tensions deriving from primogeniture rules was to limit them by law. In Islamic countries, the Holy Qur'an laid down precise rules on how to distribute inheritance, ensuring that everyone would get a fair share. In China, male heirs were privileged according to Confucianism, but the inheritance was distributed among all the heirs, although not in equal shares. India was a patchwork of different cultures, some of which practiced a relatively strict primogeniture method, but not everywhere.

Ancient Europe, instead, found a clever way to tweak the system in such a way as to make strict primogeniture possible without creating bands of penniless warriors in search of dragons to kill or scheming viragos in search of nobles to seduce. It was done by using monastic and religious orders, traditionally requiring chastity from members. Monasteries and nunneries could provide a comfortable refuge for disinherited male and female nobles who would find food, shelter, and a certain degree of comfort there. The emphasis on chastity ensured that the cloistered nobles would leave no descendants so that the firstborn heir wouldn't risk being ousted by a nephew. The system had the additional advantage that cloistered heirs could be "de-cloistered" and brought back to the inheritance line if the first-born heir were to die. An example is that of Ferdinando de' Medici (1549 -- 1609), who inherited the title of Grand Duke of Tuscany when his older brother Francesco died in 1587, leaving no legitimate male descendants. Earlier on, Ferdinando had been ordained a cardinal, and he was supposed to pursue a religious life of chastity. But when he became Grand Duke, he had no problem in "de-cardinalizing" himself, getting married, and having children. Incidentally, Francesco's death was widely attributed to poisoning. Hence, we may suspect that his younger brother had something to do with it, but we will never know for sure.

You may think that chastity was a condition that made cloistering a heavy punishment, but that was necessarily the case. In ancient times, people saw chastity as a virtue leading to Paradise in the afterlife. Besides, monasteries were not necessarily places of penance and suffering. In comparison with the ordinary life of peasants and low-class citizens, they could be seen as even luxurious. In addition, whether monks and nuns were actually chaste is a debatable question. For obvious reasons, we will never have verifiable data on this point; however, you can gain insight by reading Boccaccio's Decameron, a collection of stories published in 1349. Boccaccio pokes fun at monks and priests, giving the impression that their chastity was considered a joke by everyone, including the monks and the nuns themselves. In other sources, nunneries were accused of being run as brothels. It is perfectly possible since, just like every institution in the real world, they needed money. An example is the story of Sister Virginia Maria (born Marianna de Leyva y Marino; 1575 – 1650), who lived in Monza in northern Italy. Sister Virginia became rather infamous in her time as "The Nun of Monza" due to an affair she had with a local aristocrat, which involved the homicide of another nun who had threatened to reveal the story. Her life inspired one of the characters in Alessandro Manzoni's novel *The Betrothed* (1840).

Maybe chastity was taken more seriously in other places and times, but we may suspect that it was always more theoretical than real. In any case, the important point of cloistering was not chastity; it was making sure that monks and nuns would have no children, at least officially. If they did, the children were to be quietly disposed of. Sometimes they were killed, but, less drastically, they could be transferred to orphanages, which would ask no questions, raise them, and then reinsert them into society as "innocents." They could become regular citizens, and if they were female, the hospice would provide them with a small dowry, allowing them to marry and raise a family. The critical feature of these institutions was that all bonds with their original parents were forever severed; the innocents received a new name and would never know who their parents were. That made these hospices function as machines to turn nobles into commoners. It solved a problem that, much later, Peter Turchin defined as "Elite Overproduction" as a source of social instability.[82]

An interesting element of the orphanage scheme is that the abandoned children were often endowed with half a coin or other objects that, in principle, could allow them to be recognized by their parents, who kept the other half. You can still see a collection of these half-coins in the museum of the *Spedale degli Innocenti*, the Innocents' Hospital, in Florence, which catered for abandoned children for at least four centuries. These old objects are still moving to see today, when you think of how hard it must have been for the parents to give away their child forever. Indeed, the discovery of one's real parents is a common trope in modern and ancient fiction. Think of the myth of Oedipus, based on the dramatic discovery that Oedipus' wife was actually his mother. The film *Mamma Mia!* (2008) is a recent example.

The European primogeniture laws and customs were a clever arrangement that favored wealth concentration, which could then be turned into military power. It may have been an important factor in the competition for world domination that started with the Renaissance. This idea has been proposed, among others, by David Landes (*The Wealth and Poverty of Nations,* 1998), Timur Kuran (*The Long Divergence,* 2011), and Peter Frankopan, in *The Silk Road* (2015).

Today, monasteries and nunneries are out of fashion, and the orphanages of old are mostly closed. They are no longer necessary with the rapid decline in birth rates. Nevertheless, the problem of inheritance and distribution of wealth from one generation to another remains important. But it is an ongoing story that we will see in detail in another chapter.

Polynesia: Where Malthus Ruled

Figure 18 - The Moai, the stone statues of Rapa Nui. Image by Juan Carlos Araya Mardones. Wikimedia, Creative commons

The Rapa Nui Island, also known as Easter Island, is a poster child of catastrophists. It is often described as having suffered a major population crash caused by overexploitation of the land, a few centuries before European explorers arrived. But the real story turned out to be more complicated and nuanced than this simplified version makes it appear.

Rapa Nui was discovered by the Dutch explorer Jacob Roggeveen on Easter Day, April 5, 1722, and he and his crew were the first people from outside the island to see the famous Moai, the giant stone statues. It was a fascinating discovery, suggesting that Rapa Nui had been inhabited by an advanced civilization despite being a small island of approximately 164 square kilometers.

But when Europeans first landed there, Rapa Nui didn't host anything that could be defined as an advanced civilization. Some 3000 inhabitants lived there, clearly unable to create stone statues of the size of those that their ancestors had built. The Europeans who landed there raided the island for its only natural resource, people to be sold as slaves on the continent. Later, the island was also exploited as a pasture for sheep imported from the continent, while the surviving natives were confined to a small portion of it. After a century of disasters, a census carried out by the Chilean navy in 1892 recorded only 101 inhabitants, of whom only 12 were adult men. The Rapa Nui culture and population were on the verge of extinction.

Eventually, the islanders managed to recover, and today, some 3000 people inhabit the island, most of whom are of Rapa Nui descent. But a lot has been lost. Nowadays, the Rapa Nui language is spoken by no more than a few hundred people, and the ancient culture has been so forgotten that we don't even know the original name of the Island. ("Rapa Nui" is a modern term.) The ancient Rapa Nui people had a written language known as Rongorongo, one of the most enigmatic and undeciphered writing systems in the world. Only 27 Rongorongo tablets survive today, and their meaning is forever lost.

The mystery shrouding Rapa Nui and its statues stimulated the fantasy of Western explorers. Thor Heyerdahl (1914 – 2002), a Norwegian ethnographer, was one of the first scientists to explore the island in the 20th century. He mentioned the possibility that the Islanders had destroyed themselves by overexploitation, and he proposed that the final stages of the collapse involved wars and perhaps even cannibalism. The concept of ecological overexploitation on Rapa Nui was later discussed by John Flenley and Paul Bahn in their 1992 book *Easter Island* and then popularized in the 1994 film *Rapa-Nui*. The idea was also described by Jared Diamond in his 2005 book *Collapse: How Societies Choose to Fail or Succeed*.

The common thesis of these interpretations was that the inhabitants of Rapa Nui deforested the land to leave space for agricultural production and also for moving moai statues on tree trunks used as rollers. This led to soil erosion and loss of resources, ultimately resulting in an excess population. Without sufficient food, the population collapsed among conflicts and cultural upheaval; a Malthusian collapse if ever there was one. Therefore, the island could be seen as a model of the larger island, Planet Earth, offering us a vision of our bleak future.

It is always dangerous to theorize on the basis of a few data, and the theories on Rapa Nui's population were based on almost no data. It is not surprising that they turned out to be only partially correct, if not completely wrong. The idea that the islanders had cut down the trees was contested, proposing that the rats introduced by European visitors had caused the deforestation by eating the seeds of the island's palm trees. That was another case of a hypothesis based on nearly no data. Recent studies show that the ancient Rapa Nui trees were cut using axes. Rats surely couldn't do that.[83]

So, humans did destroy Rapa Nui's forests, but did that cause a population collapse? That was mainly a hypothesis based on the idea that the modern population of the island was too small to be able to create statues of the size of the Moai. But studies based on genomic data revealed no evidence of a "genetic bottleneck" that would indicate a drastic reduction in population before the arrival of Europeans.[84] Evidently, the Rapa Nui people were able to maintain food production at a level sufficient to avoid a major population collapse, despite having deforested their island. If you think about that, indeed, there is no obvious reason why deforestation should have caused famine and depopulation. Rapa Nui is a minuscule green spot in the vast Pacific Ocean. Its climate is tropical, characterized by high humidity. The volcanic mountain Maunga Terevaka reaches 505 meters in height, and it affects the climate by creating rain out of wet winds that blow over the island. Whether Rapa Nui is forested or not, the weather pattern remains approximately the same. Of course, without trees, there is always a risk of soil erosion, but we have archaeological traces showing that the islanders took care of their cultivations by making rock barriers against soil loss. The lack of trees must have made it difficult for the islanders to cook their main agricultural product: sweet potatoes. But, again, it seems that the island had sufficient shrubs and dry grass to make this possible.

The results of the genome examination of Rapa Nui inhabitants led to an active phase of "Diamond-bashing," in which critics accused him of twisting data to suit his preconceived ideas. However, as usual, excessive leaning in one direction or another doesn't lead you closer to the truth. The Rapa Nui islanders may not have suffered a population collapse, but at the very minimum, they saw an abrupt cultural change. They abruptly stopped making the Moai, leaving some of them unfinished in the caves where they manufactured them.

One fashionable theory popular up to not long ago was that deforestation forced the islanders to stop manufacturing the Moais due to the lack of wooden logs as rollers to transport them. But this idea is nowadays largely abandoned. It is believed that the islanders "walked" the statues to their places by a controlled rocking method while pulling them with ropes. It may be more likely that the cultural change was caused by the loss of the capability to build boats.

Despite its isolation, Rapa Nui probably maintained contacts with other Pacific Islands and maybe with the South American coasts. There is robust genetic evidence for intermixing between the Rapa Nui people and Native South American populations, occurring well before European contact in 1722. This admixture is estimated at about 8–10%. It is remarkable how the Islanders covered a distance of nearly 4,000 km from Rapa Nui, and then back; a distance equivalent to that from London to New York. But the Polynesian double canoes were sophisticated ocean vessels that could make the trip in a month or so, carrying enough water and food to sustain the crew during the voyage.[85] Losing the supply of wood necessary to make canoes locked the islanders forever on their island, and that may have led to a loss of cultural and genetic exchanges, weakening the Rapa Nui culture and leading them to abandon some of their traditions.

But if Rapa Nui cannot be a model for global population collapse, there are cases of Polynesian islands where resource overexploitation did lead to collapse, as Diamond himself reports in his book. One fascinating case is that of the Pitcairn Islands, a group of four volcanic islands located in the southern Pacific Ocean. Two of them, Pitcairn and Henderson, were inhabited for a few centuries by Polynesians who arrived there probably during the 11th Century CE from Mangareva Island, 540 kilometers to the northwest. All these islands are much smaller than Rapa Nui. Mangareva has an area of approximately 15 square kilometers, Pitcairn Island has an area of 4.6 square kilometers, and Henderson Island has an area of 37.3 square kilometers. Living on such small islands must have posed enormous survival problems to the inhabitants. Even though they could rely on fishing for their food, they still needed wood for their canoes and stone for their tools. Their survival appears to have been based on trade. Mangareva supplied oyster shells for fishhooks and ornaments, and also wood for canoes. Pitcairn supplied basalt for adzes and obsidian for cutting tools, while Henderson supplied turtle and bird eggs, bird feathers, and fish.

Something went wrong during the 15th century, when trade between these three islands ceased. Most likely, it was the result of the deforestation of Mangareva, the only one of the three islands that could supply the timber needed to make canoes. Indeed, when Europeans arrived in Mangareva in the 18th century, they found that the inhabitants had rafts but no canoes.[86] It seems that the three islands, Pitcairn, Henderson, and Mangareva, could exist only by supplying each other with commodities that couldn't be produced locally. But only Mangareva could host a human population in isolation. Pitcairn couldn't produce sufficient food. Henderson, could produce seafood, fish, and bird eggs, but it couldn't produce stone for tools and wood for canoes. In both cases, the populations died out or left. Pitcairn was resettled in 1790 by nine of the mutineers from the British ship HMS Bounty, who landed there along with six Tahitian men, 11 Tahitian women, and a baby girl. They survived in precarious conditions up to the present day, reaching a maximum of more than 200 inhabitants during the 19th century. Today, it is reported that there are only 47 permanent inhabitants, and the island may see a second human extinction in the near future.

So, although deforestation didn't cause a population collapse on Rapa Nui, it did on smaller Pacific islands. Apparently, there is a critical size that makes it impossible for the inhabitants of an island to survive on local resources only. This limit depends on the geology of the island and its fertility. Evidently, the 5 square km of Pitcairn and the 37 square km of Henderson are too small for a permanent settlement. For these islands, the Malthusian limit was effectively zero, and the population had to adapt to that limit, reaching zero inhabitants. Instead, the 164 square km of Rapa Nui were enough to sustain a human population in isolation, at least for several centuries.

The size of Earth's land surface is much larger than that of any island in the Pacific Ocean, and that would seem to make human extinction in the near future unlikely. Indeed, the Earth's ecosystem has survived for more than three billion years, and modern humans have been living on Earth for a few hundred thousand years. But that doesn't mean we cannot destroy Earth's ecosystem to a level that would make its Malthusian limit for us equal to zero. Maybe we could move to Space colonies or to Mars, as Elon Musk is proposing nowadays. But these colonies might be too small, too, to survive on local resources only, and the colonists might face the same fate as the inhabitants of Pitcairn.

Greenland: Colonizing a Remote Planet

The saga of the Norse colonization of Greenland is another small-scale example of what could be humankind's destiny at the global level. A remote and forbidding land that was nevertheless colonized and inhabited for a few centuries, until the colonists had to give up and leave.

Europeans sighted Greenland for the first time when a Norseman named Gunnbjörn Ulfsson sailed in the vicinity in the late 10th century CE. The first recorded landing was by Erik the Red, an Icelandic Viking who was exiled for murder around 982 CE and was seeking a haven for himself and his followers. Erik the Red explored the southwestern coast of the place he had found and named it "Greenland." The name was a little misplaced for a land that, then as now, was mostly covered with a thick ice layer. But it is also true that the region where the Norse landed and settled was covered with green grass at that time, just as it is today. Greenland's glaciers are not visible from there, being tens of kilometers inland, and the Norse may not even have known that they existed.

Figure 19 - A modern reproduction of Thjodhild's church in Greenland. It gives some idea of the environment in which the Norse lived in Greenland. Image by Brattachur, Wikimedia, Public Domain

Erik the Red returned to Iceland, and then he led a larger expedition to Greenland in 985 CE. The Norse established two main colonies: the Eastern Settlement, located near present-day Qaqortoq, and the Western Settlement, situated near present-day Nuuk. These names are confusing: both settlements were in Western Greenland, and the "Eastern" Settlement was actually North of the "Western" one. In any case, the Norse settled in the only large ice-free region of Greenland.

The settlers created colonies that had all the characteristics of Norwegian society in Europe. They engaged in a combination of farming, hunting, and fishing, and they brought with them the whole menagerie of domestic animals they had in Europe: cattle, sheep, pigs, goats, and even horses. They also undertook expeditions to the North, where they could hunt walruses for ivory, furs, and meat. That was a vital part of Greenland's economy. Exporting ivory and furs to Europe made it possible for Greenlanders to import supplies of the resources that Greenland couldn't produce, such as wood for ships and iron for tools and weapons. The Greenlanders even traveled to Labrador, on the North American continent, probably to get timber. For some reason, though, they never settled there. If they had, they might have colonized North America much before the Pilgrim Fathers arrived there in 1620. In that case, this book would have been written in Old Norse, or maybe in Iroquoian.

The Norse colony in Greenland thrived for a few centuries, a fascinating chapter in history. From an initial settlement of probably just a few hundred people in the late 10th century, the population is estimated to have grown to about 1500 by the 12th century. At its peak, during the 13th century, the population reached between 2,000 and 2,500 people, with some estimates suggesting even 5,000.

The decline of the colony began in the mid-13th century and accelerated during the 14th century. By the early 15th century, the Western Settlement had collapsed, with the survivors probably taking refuge in the Eastern Settlement. By the mid-15th century, the Eastern Settlement was also abandoned. Some of the inhabitants may have managed to return to Europe, but most of them died, leaving no descendants. In Europe, nobody knew about the fate of this remote outpost. When Norwegian missionary Hans Egede reached Greenland again, in 1721, he found Inuit people living there, but no descendants of the old Norse settlements.

What happened to the Greenlanders that wiped them out of history? The "Little Ice Age," the cooling trend of early modern times in the Northern Hemisphere is often seen as the culprit. But it is not a satisfactory explanation since it overlaps only marginally with the timeline of the Norse settlements. When the Little Ice Age hit really hard, during the 17th century, the Greenlanders had disappeared more than one century before. We will probably never know exactly what happened, but it may be that it was the result of Shakespeare's Principle that "troubles come in battalions," a sentence pronounced by Polonius in Hamlet. The Norse colony was hit by a series of problems that reinforced each other and, eventually, destroyed it. The main one may have been the same that doomed the inhabitants of the Pitcairn Islands in the Pacific Ocean: the collapse of long-range trade.

The European Great Famine of the early 14th century and the Black Death of a few decades later coincide with the start of the decline of the Greenland colonies. With the European economy in shambles, the trade of luxury items with Greenland dwindled. Europeans couldn't afford anymore the walrus ivory and fur that they used to appreciate. That didn't necessarily mean that the people of Greenland were to starve. There is no evidence that they imported food from Europe, and it is clear that they relied mostly on local food sources for their subsistence. The problem was that Greenland's economy was too small to afford iron smelting from iron ore, and there is evidence that it fully relied on Europe for iron nails and tools. In addition, there are no suitable trees for building boats in Greenland. That could perhaps have been overcome by getting timber from Labrador, and the Greenlanders may have attempted to do just that. But transporting wood from Labrador to Greenland required large boats, and no nails meant no boats.

You see how problems came in battalions. Without boats, the Norse couldn't get timber from Labrador. And they couldn't even organize hunting expeditions to the North for walrus furs and ivory. Without walrus furs and ivory, they couldn't pay for importing nails from Europe. And without nails, they couldn't make boats. A vicious circle, if there ever was one. The result must have been that their trade-based economy collapsed. There is some archaeological evidence of severe nutritional stress for them, although there is no evidence of a major wave of famine-generated deaths. In any case, they may well have concluded that it made no sense for them to stay in a gloomy and cold place that offered them nothing but a poor diet and poverty. So, most of them decided to leave when they still had a chance to. Maybe they saw themselves not as "Greenlanders" but as an outpost of their homeland, where they were always planning to return one day or another.

The disappearance of the Norse settlements in Greenland gives us a chance to note how historical events can be completely misunderstood in our times. In the book *Collapse* (2005), Jared Diamond pokes fun at the settlers for not attempting to eat fish to survive, noting that fish are abundant in the seas around Greenland. He attributes its absence in the Greenlanders' diet to superstition, "a taboo against eating fish," maybe because some of the early settlers had suffered from food poisoning by eating spoiled fish. If the Greenlanders had been smarter, Diamond implies, they would have survived.

I am not a fan of "Diamond-bashing," but in this case, it is true that he completely missed the point. As usual, the absence of data doesn't mean that something doesn't exist. Some archaeological remains of fish bones in Greenland's settlements have been found recently, and, in any case, it would have been weird for the Greenlanders to have developed a taboo against eating fish. They came from Norway, where everyone eats fish, and a lot of it; no taboo whatsoever exists about that. And Greenland's colony was never completely isolated from the mainland.

The problem is not about taboos. It is that, as we already discussed in this book, no society in history has ever survived on fish only. It is because of the unreliability of the yield of fishing and the difficulties of conserving fish for a long time. In Europe, Norwegians had developed effective fish-drying techniques, but they could never survive on fish alone. In any case, drying fish Norway-style was possible only in the dry climate of Norway. In the wetter and colder polar climate of Greenland, drying fish is nearly impossible. Note also that Greenland had no salt resources, so fish couldn't be preserved by salting. In short, the Greenlanders simply couldn't survive on fishing only.

We may speculate on whether the Norse could have adopted the lifestyle of the Inuit, who, at that time, were living in Northern Greenland. The Inuit were stone-age hunters who had taken to the sea, and they had developed a remarkable set of technologies that allowed them to survive in extreme polar conditions. They used light boats called *kayaks* and *umiaks*, made with a frame of bone or wood, covered with animal skins, without the need for timber to make planks, or of nails to keep them together. They hunted and fished with stone-tipped spears, and they could conserve meat and fat for long periods in their cold climate. Contrary to the common belief, Greenland's Inuit did not normally live in the classic snow-block igloos typical of the Canadian Arctic Region. They tended to live in semi-subterranean houses with thick walls and roofs made of stone and turf (grass/soil blocks).

The Greenland Norse never adopted these techniques. Probably, they didn't have enough time to learn, and it appears that their contacts with the Inuit were sporadic and often unfriendly. In addition, the lifestyles of the Norse and of the Inuit were completely different. For instance, the Norse needed large enclosed spaces for their livestock, and that was nearly impossible to obtain with the small Inuit houses. We may wonder whether some of the last Norse of Greenland could have been adopted by the Inuit and survived the collapse of their colony. It is not impossible, but modern studies have found no genetic evidence of mixing between the two populations during Medieval times. When the last Norse left Greenland, or died, the colony disappeared completely

The story of the Norse colony in Greenland may tell us something about the current idea of colonizing Mars, proposed by Elon Musk, and also a staple idea of Science Fiction. In many ways, it would be a task similar to the colonization of Greenland: a group of adventurous people settles in a remote and cold place with the idea of living away from the mainland. But, just like the Norse Greenland colony, a Martian colony could survive and become self-sufficient only if it could engage in a reciprocally advantageous trade with the mother planet, Earth. Earthlings wouldn't be happy to supply their colony for a long time without getting something back in exchange. But what products could a Martian colony ship back to Earth? For sure, there are no walruses on Mars that could be hunted to ship walrus ivory to Earth. Maybe Mars could develop a product that can be manufactured only on Mars, but what? "Unobtanium," produced on Planet Pandora in the *Avatar* movie series, is a purely fictional concept. Nevertheless, Mars might have some abundant mineral resources generated during its long geological history and then frozen because no more geological evolution has taken place there during the past 3 billion years or so. The cost of shipping minerals from other planets to Earth is not necessarily insurmountable. Space travel is prohibitively expensive when it involves human beings who cannot withstand high accelerations and need complex and expensive life support systems. But interplanetary shipping could be done using electromagnetic catapults, propelling a payload pod to high speeds by means of electromagnetic fields.

Gerard O'Neill, the proposer of large-scale space settlements, is credited with having pioneered the idea of magnetic catapults, or "mass drivers," with his book *The High Frontier* (1976). But the first mention of magnetic catapults launching payloads into space may have been by Robert A. Heinlein with his *The Moon is a Harsh Mistress* (1966). It was a science fiction novel in which he imagined that the human colonies on the Moon could produce grain in underground cultivations and ship it to Earth to feed a population that had become so large that it couldn't rely solely on Earth's agriculture.

One of these catapults works very much like a maglev train, except that it operates under a vacuum. Hence, it can reach much higher speeds than any vehicle running on Earth's surface. Several experimental tests have been performed, and the concept appears feasible, although so far, no payload has been sent to orbit from Earth by one of these machines. Besides, a human being shot to space by an electromagnetic catapult would be squashed flat to something thinner than when Wile E. Coyote is run over by a road roller in cartoons. But such high accelerations would do no damage to bulk commodities such as metals.

So, would it be possible to move large numbers of people to other planets to overcome the limits of Planet Earth? In the 1978 book *Doomsday Has Been Cancelled* by J. Peter Vajk and Carolyn Henson, the authors developed an optimistic rebuttal to *The Limits to Growth*, proposing that space colonization—through space habitats, mass drivers, and solar power—can overcome Earth's resource and population constraints. By expanding humanity's resource base and living space, the authors argued that the collapse scenarios could be delayed or avoided.Not impossible, but at present very unlikely, to say the least. As things stand, at best, a Martian colony could operate like the modern settlements in Antarctica. They are scientific outposts where people live only for limited periods, and where nobody would like to live for their whole life. To date, only 11 people are reported to have been born in Antarctica. And note that Antarctica is not as cold, barren, and distant as Mars.

The human adventure on Earth shows that traveling and colonizing remote lands is a difficult and risk-filled enterprise. But, in many historical cases, brave and resilient explorers could overcome all difficulties and settle in remote and barren lands. We don't know if, in the future, it will be possible for humans to colonize space or other planets. But our silicon children, robots, don't need oxygen or special conditions of temperature. In the future, they may expand where no human beings could survive.

Ireland: The Great Famine

Ireland's great famine, known as *An Gorta Mór* in Gaelic, is perhaps the best-known example of how famine can cause the collapse of a human population. It began in 1845 and, in a few years, about half of the Irish population had disappeared, directly killed by starvation, or by sickness, or forced to emigrate. It was a shocking event when it occurred, and it is still shocking for us after nearly two centuries. How could that happen? Ireland in the 19th century was part of a world in which there were, in principle, resources and technologies that could have avoided the disaster or at least mitigated it. But, as it often happens, disasters occur because people do too little, too late to cope with them.

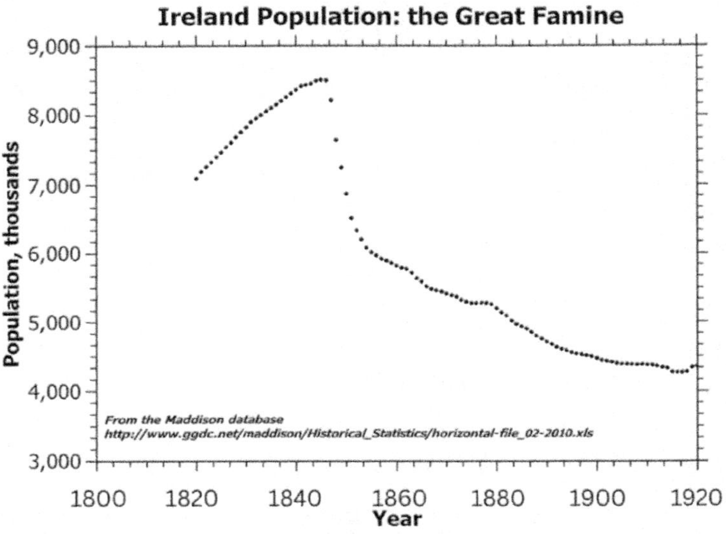

Figure 20 – Ireland's population before and after the Great Famine. Data from the Maddison Foundation

In ancient times, Ireland was a sparsely populated land covered with thick forests. It was too poor and too far away to be worth conquering, and so the Roman legionnaires never set foot there in force. With the fall of the Roman Empire, Ireland was shielded by its isolation from the wars and invasions that occurred in Europe. As a relatively unscathed region, it played an important cultural role in early medieval Europe by saving many ancient manuscripts, conserved and copied by Irish monks. But, as continental Europe recovered from the collapse, it became clear that Ireland was too small and too weak to compete with the new European powers. Of these, England was the closest and the most dangerous one.

There were several waves of invasion of Ireland by English armies. The final one was led by Oliver Cromwell and ended in 1653 with the complete conquest of Ireland and the collapse of the Irish social structure. The ancient Irish nobility was exterminated or sent into exile. The surviving Irish lost all their lands, had little or no civil rights, could not carry weapons, and couldn't vote or have representatives. All economic activities fell into the hands of the English rulers, while the Irish were relegated to the roles of peasants or workers. Ireland was the first colony of what would later be called the British Empire, even though it was never formally declared one.

According to the data by Cullen in *An Economic History of Ireland Since 1660* (1972), after the conquest, Ireland quickly became an economic asset and a food provider for England. So much that it was called "England's larder," or "England's breadbasket." It could produce meat, cheese, butter, honey, and other high-quality food for the English, while the Irish had to content themselves with potatoes. Yet, despite intermittent hard times, an Irish family could get enough food to survive from a relatively small patch of land by cultivating potatoes and cooking them using local peat as fuel. When they were not working on their potatoes, they worked for their landlord in order to pay the rent. For the landlords, it was the perfect arrangement: they had manpower for free. For the Irish, it was far less than perfect, but at least they could survive. They lived in mud cabins or cottages with walls made of mud, clay, or sods (stacked turf). Floors were often just packed earth, windows were small or absent, and of course, there was nothing like running water and toilets. These cabins had one room shared by families, and often also by livestock. Furnishing consisted of a few stools and straw beds. Several commenters of the time mentioned that overseas "savages" lived better than the Irish.

But the problem for the Irish was not the appalling conditions of their daily life. It was that relying on a single crop is risky, and intermittent famines were a normal feature in Ireland. It was one of these periodic famines that led Jonathan Swift to write his pamphlet *A Modest Proposal*, (1729), where he satirically proposed to solve the problem of hunger in Ireland by having the Irish eat their children. He preceded by two and a half centuries the idea of *Soylent Green*, fictional food made from human flesh, that would appear in a 1973 movie with that title.

We don't know whether the rapid growth of the Irish population before the Great Famine was actively encouraged by the English landlords in order to have more manpower available. It may also be that the Irish reasoned as all peasants in the world do: more arms in the family, more wealth. Perhaps they also dreamed that their growing numbers would one day enable them to rid themselves of their hated English masters. Whatever the case, Ireland experienced a real explosion in population during the British rule. There had been fewer than two million people in Ireland at the time of the English conquest. By 1840, the historical population maximum had been reached: more than 8 million.

At the same time, making room for food production spelled the end of the Irish forests. We don't have to think that before the English conquest Ireland was still covered with the lush forests of the time of the mythical hero Cuchulainn. But the data reported by Michael Williams in his *Deforesting the Earth* (2003) show that Ireland in the 17th century still maintained about 12% of the land covered with trees, significantly more than most European countries of that time. The deforestation of Ireland is described in Eileen McCracken's book *The Irish Woods since Tudor times* (1971), in which we read that Ireland ceased to be an exporter of wood and became an importer by the late 18th century. Not just forests were destroyed, but also the species that inhabited them. Squirrels and deer went extinct, to be reintroduced only in the 20th century. The last wolf of Ireland is reported to have been shot in 1770. Travelers in Ireland at that time reported a treeless, bleak landscape. The loss of biodiversity and the erosion of the soil were elements of a disaster waiting to take place. And it did.

In 1845, the leaves of the Irish potato plants started turning black and curled, then rotted. Nobody knew that it was because of a fungal infestation, but that mattered little; the potato harvest was ruined, and at the end of the year, the Great Famine began. About one million people died of starvation in a few years. Another half a million died of diseases related to starvation. Many tried to escape and, in many cases, they left Ireland for the United States on ships that were totally inadequate in terms of hygienic conditions and lacked sufficient food and water. These vessels were termed "coffin ships" and caused the death of an untold number of Irish emigrants. Eventually, the famine caused Ireland to lose about half of its population, four million people.

Remarkably, even during the worst years of the famine, Ireland continued to produce food to export to England. That generated the legend of the "Irish Holocaust," which blames the English for having willingly created the famine to get rid of a large number of their Irish subjects. But if the British wanted to get rid of the Irish, they could have used the method that Cromwell had successfully used a couple of centuries before: bullets. Since the Irish had no army, no weapons, and no training in military matters, they couldn't possibly have resisted the extermination treatment that the US government was inflicting on the Native Americans at the same time. More likely, the British government had no interest in exterminating the Irish, a population that was already working for them at essentially zero cost. The British poor, then, had much better prospects of emigrating overseas than of becoming even poorer by settling in Ireland, where their only perspective was a diet of potatoes and a life in mud cabins.

Unfortunately, if the British had no interest in exterminating the Irish, they had no interest in their survival, either. They didn't care about people whom they considered unruly, unskilled, and speaking a barbarous language that nobody could understand. Nevertheless, when the famine came, the British government initially provided some relief. The government headed by Robert Peel (1841–1846) repealed the Corn Laws that imposed a tax on imported grains. Indian maize was purchased specifically to be sent to Ireland, and help was provided in the form of "soup kitchens." Most likely, what these kitchens served wasn't exactly gourmet food, but they saved hundreds of thousands of people during the initial phases of the famine. But these attempts were discontinued by the government of Lord John Russell, who succeeded Peel in June 1846. Russell's government reinstated the corn laws and dismantled the soup kitchens, leaving the Irish to paddle their own coracles, so to speak. And that created the disaster.

So, the British didn't plan the Irish genocide, but they made it possible by omission. They chose to let the Irish starve and die. Why did they do that? After all, other European countries, such as the Netherlands and France, behaved differently, and the potato blight didn't cause mass starvation because emergency food relief was provided to the poor. If we go back to the debate in the English documents and newspapers of the time, we see how the Irish are often described as an inferior race and culture; lazy and good-for-nothing people who created by themselves the disaster that befell them. Jonah Barrington, English landlord, describes the Irish peasant of his times (late 18th century) as a mixture of the troll and the hobbit; sometimes dangerous, sometimes funny, never fully human. Benjamin Disraeli (who became prime minister after the Irish crisis was over) is reported to have said, "The Irish hate our order, our civilization, our enterprising industry, our pure religion" (we read this sentence on page 6 of Cahill's book *How the Irish Saved Civilization* (1995). Apart from the insults, the general idea was that helping the Irish would have done more harm than good by creating a persistent dependency. Charles Trevelyan (1807-1886), assistant secretary to the treasury, expressed the concept as "If the Irish once find out that there are any circumstances in which they can get free government grants, we shall have a system of mendicancy such as the world never knew."

In the end, the Irish population collapse was a good illustration of how complex systems collapse: they never do that as the result of a single factor, but as many factors converging, all pushing in the same direction. And when that happens, the result is not just decline, but collapse – it is the Seneca Effect. The Irish society didn't collapse just because a fungal infection hit the potato crops. It collapsed because the infection triggered a series of consequences that were built into the structure of the Irish society. As soon as the infection unbalanced the system a little, all these factors started to play their role: overpopulation, lack of road transportation, lack of good ports in the Southern region, absence of large scale social networks, religious hatred between different sections of society, and, worst of all, a central government which didn't care about the welfare of its Irish subjects.

So, what had to happen, happened. Can it happen again on a larger scale in our modern world? Earth is a large island in comparison to Ireland, but it is still an island. And we cannot discount the possibility that events similar to the *An Gorta Mor* will unfold on a much larger scale in the future.

North America. Depopulation by Culling

During the 19th century, the largest genocide in the history of humankind took place in North America. It was the culminating phase of a cycle that started when Columbus reached the New World in 1492. Already, the first contact between Europeans and Natives resulted in local conflicts, but the early centuries of interactions also involved considerable cultural interchanges and reciprocal influences.

What doomed the Natives had a specific name: the Industrial Revolution.

It started in the United States in the 18th century, with the first coal mine established in 1748 in Virginia. By the late 18th century, the Eastern coast of the US was already in a full industrialization phase with textile mills, forges, and the production of weapons and ammunition. The American Revolutionary War (1775-1783) showed that the US industry could support a war effort to match an established world power such as Britain.

The booming economy of the East Coast led to a gradual process of expansion of the Whites. The Natives tried to resist by military means, initially with some success. At the Battle of the Wabash in 1791, the Natives were able to inflict on the Whites possibly the worst defeat ever experienced by the US army in its history. But, from then on, it was all downhill for the Natives, who were able to defeat the US forces only occasionally in local skirmishes. That was what you would expect when a society that was at a stone age technology level faced a modern industrial power.

The 19th century saw the expulsion of the Natives from their lands, and their gradual extermination. The Indian Removal Act (1830) under President Andrew Jackson led to the "Trail of Tears" when eastern tribes (e.g., Cherokee, Creek, Choctaw, Chickasaw, Seminole) were forced to relocate to lands west of the Mississippi River. Over 60,000 people marched along the Trail of Tears. Many died along the road of disease or starvation. Many others died or were killed later as they couldn't adapt to the new lands. The Seminole population in Florida was nearly completely exterminated between 1816 and 1858. In 1832, the Sauk and Fox tribes were pushed out of Illinois and Wisconsin, and most of them died in an equivalent of a new Trail of Tears. Later, around 1876-1877, it was the time of the Lakota, Cheyenne, and Arapaho, led by figures like Sitting Bull and Crazy Horse. It was the time of the last Native victory at the Little Bighorn (1876), but that had no effect on the outcome of the war. The surviving Natives were forced onto reservations chosen among unproductive lands where they could only live in conditions of extreme poverty. The massacre of Wounded Knee was the last event of some relevance in this war of extermination.

It is estimated that the Native American population in the U.S. dropped from an estimated 5–15 million (pre-Columbian) to under 250,000 by 1900. Estimates vary about the fraction of the Native American population exterminated; some are as high as 98%, others suggest a less drastic "culling" of about 70%-80%. At that point, Native Americans controlled less than 2% of their original territories in the U.S. Their cultures were thoroughly suppressed, and very little remains of them, with only a fraction of US citizens of Native descent still speaking the old languages.

It was one of the largest, most thorough, and most ruthless exterminations ever reported in human history, although maybe not so ruthless as the extermination of the Amalekites, a population living in Palestine, carried out by the Jews in biblical times. According to the Bible, God ordered the Jews to kill not only the Amalekite men, but also the women and the children, and even their livestock. Maybe this extermination was not so thorough as God had commanded it to be, but we have plenty of data about more recent ones. We may consider the Belgian Atrocities in the Congo Free State (1885–1908), the Herero and Nama Genocide by the Germans in Africa (1904–1908), the Armenian extermination by the Ottoman Empire (1915–1923), the holocaust of the Jews and other groups in Germany (1941-1945), the Rwandan Genocide (1994) and others ongoing right now. But the extermination of the Native Americans preceded all these cases, and was much larger in terms of absolute numbers and thoroughness.

As I discuss in my book *Exterminations* (2024), when certain conditions are satisfied, extermination is nearly unavoidable. It happens when a subgroup is 1) easily identifiable, 2) unable to provide strong military resistance, and 3) provides monetary gains for the exterminators. In addition, an extermination needs to be accompanied by a certain degree of hate. Killing a person in cold blood is possible only when the killer is somehow convinced that it is proper and necessary to do so. That's the job of modern propaganda, which usually succeeds in the task.

When these conditions hold, eliminating people who don't belong to one's country or ethnic group has never been especially difficult in history. President Andrew Jackson (1767 – 1845) is reported to have used the phrase "Nits make lice" to justify the extermination of Native women and children. Then, of course, talking about exterminators, Adolf Hitler is supposed to be the one who rules them all. However, we have to note that he took inspiration from the way White Americans treated the American Natives.

In addition to the extermination of the Jews and other minorities, Hitler's Nazi government is credited with having planned a culling that would have been the largest in history if it had been carried out to completion. It was the *Generalplan Ost* (General Plan for the East), also known as the Ostplan, which was developed in 1942 during World War II. The proposers of the plan had in mind the concept of overpopulation and thought that the Slavs living in Eastern Europe were a "surplus" that could be conveniently removed to leave space for the superior German race. They imagined that, after having defeated the Soviet Union, the lands West of the Urals were to be occupied by 6 to 12 million German farmer-warriors, the *Herrenvolk* (master race), while several tens of millions of ethnic Slavs were to be killed in place, or pushed on the other side of the Urals to die of starvation. It would have been an operation very much like the "Trail of Tears" of the Native Americans of 1830, but on a much larger scale.

The example of the *Ostplan* is one of the several large-scale culling operations that were imagined but, fortunately, never put into practice. Unfortunately, though, in our crowded world, the idea of the wholesale elimination of disliked people remains popular. There are plenty of possibilities for governments to engage in major culling operations. They are already demonstrating that they can do that perfectly well by engaging in a series of major wars when we desperately need to conserve energy and resources to face the challenges ahead. But this is the way the human mind works.

Japan: Stability During the Edo Period

During the Edo period (1603-1868), Japan was nearly completely isolated from the rest of the world, and the Japanese had to survive on no more than what they could produce in their islands. This period can be seen as a test bed for Malthus' ideas. There was a limited territory available and, as a consequence, a limit to the food supply that could be produced. How would the Japanese population adapt?

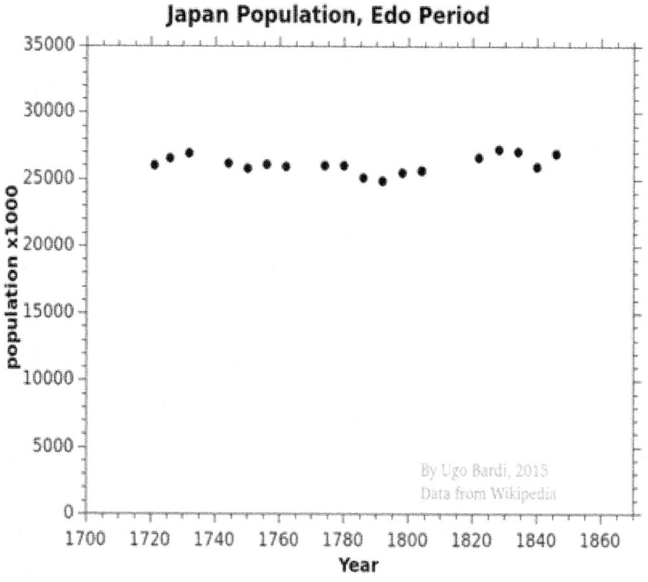

Figure 21. The Japanese population during the Edo Period

Apparently, they succeeded at the task. The population of Japan remained nearly constant, around 27-28 million inhabitants, without suffering large famines. It was a remarkable success considering how another island on the opposite side of Eurasia, Ireland, saw a succession of tragic famines.

The Japanese of the 19th century knew the work of Malthus, but we have no evidence that their government relied on Malthus' theory. Their approach was mostly pragmatic, influenced by the neo-Confucian views from China, which emphasized harmony between humans and nature. Since Japan was an island, the Japanese didn't face immediate threats from their neighbors, nor were they in a condition to attack them. So, the Government had no interest in having expendable cannon fodder, but rather in having a stable population. Stabilization was obtained mainly by "bottom-up" strategies enacted by single women or families. The Japanese used various contraception methods and, only when that was not enough, infanticide, called *Mabiki* ("thinning," a term borrowed from rice farming). Occasional famines were caused by bad weather or volcanic eruptions, but they never had more than a limited impact on the population. It was a remarkably successful story that generated the blossoming of a sophisticated civilization, of which we still admire art and poetry.

Things changed when Japan ceased to be an isolated land because of new technological developments and the economic expansion of the West. In 1854, a small fleet of American warships arrived in the Bay of Tokyo. They were called the "Black Ships" by the Japanese. The admiral of the fleet, Commodore Perry, forced the Shogunate government to open the country to foreign commerce. It was the start of a profound transformation of Japanese society, and one of the results was a population boom that continued up to the early 2000s. You see in the figure a plot of the population increase that went from less than 30 million in 1850 to a peak of 128 million people; an increase of more than a factor of four.

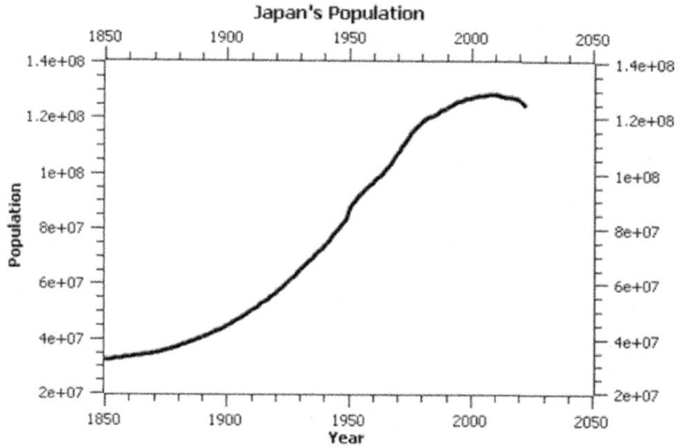

Figure 22 – Historical Japan's population. Data from Our World in Data

Today, the Japanese population is approximately 123 million, slightly lower than what it was at the peak in 2010. It is still a large number, and the problem is clear. During the Edo Period, the Japanese agriculture could feed no more than 30 million people. How can it feed more than a hundred million people today? Mainly, it is because of artificial fertilizers and other innovations; nevertheless, more than 60% of the calories consumed in Japan today come from imported food. Hence, in terms of food supply, Japan is one of the most import-dependent nations among developed countries. At present, the Japanese economy is healthy enough to be able to pay for these imports. Still, it is unsettling to think of what could happen to Japan if the economy were to decline while the world is facing a food shortage.

Fortunately, the Japanese population has started declining, prefiguring a possible return to conditions of self-sufficiency. Nevertheless, with the current annual degrowth rate of -0.54%, it would take more than two centuries for the Japanese population to return to the level of the Edo period of approximately 25 million people. But the decline is accelerating, so that Japan may return to sustainable conditions in a much shorter time. But would that be possible fast enough and, at the same time, without a traumatic collapse?

Recently, public awareness of the risk of food shortages has been growing in Japan, and the government is launching campaigns promoting more domestic production, while some citizens are stockpiling food. Surveys show rising anxiety, especially post-COVID and amid recent global instability. So far, it is nothing that could be defined as panic; it is more like a simmering unease. Would Japan face the same destiny as Ireland or some Polynesian islands? We cannot say, but we can only hope that the Japanese can manage a non-violent population decline that will lead them to a safer situation.

China: Birth Control and the One-Child Policy

China is the only country in the world that engaged in a birth control policy that involved direct government intervention to limit the number of children that a couple could have. A series of regulations were enacted starting in 1979-1980 under the name of the "One Child" policy. Having more than one child was not a crime, but it was treated as an administrative violation that carried financial penalties. The regulations remained in force until 2015, although they were progressively relaxed from the early 2000s.The Chinese birth control policy is normally described as a success for the purposes for which it was enacted, that is, reducing population growth. Indeed, the Chinese fertility rate rapidly declined from the 1970s onward. However, in the West, this policy is generally criticized and condemned as a mark of dictatorship and oppression. The debate is clouded with thick layers of propaganda, but the real story may be more complex and nuanced than the way it is normally presented. Why exactly did the Chinese government embark on this policy? And was it really effective? Understanding this matter requires revisiting the history of China and its population.

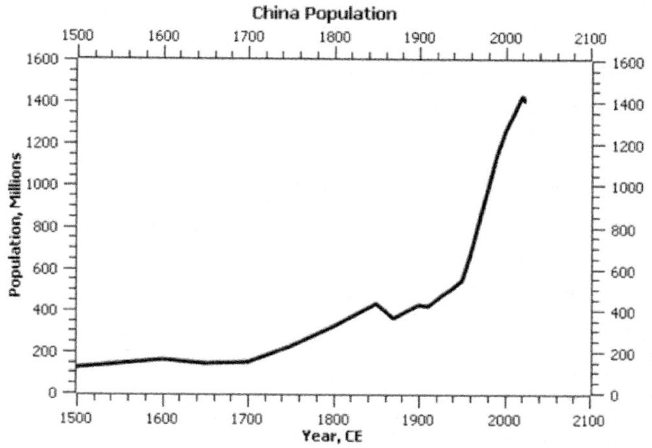

Figure 23 - The Chinese Population during the past 500 years. Data from Wikipedia

In ancient times, during the first unified Chinese empire under the Qin Dynasty (221-206 BCE), China's population was around 20-40 million people. Through the various dynasties that followed, the population grew and declined during alternating periods of peace and turmoil, but it remained always below 200 million. Things changed with the Qing Dynasty (1644-1912) when the population started growing, reaching more than 400 million by the mid-19th century.

This burst of population growth can be explained by the role that China played during the age of expansion of the European powers. The Industrial Revolution started in the mid-18th century, and it gave Europeans a technological and economic advantage that allowed them to embark on the conquest of nearly the whole world.

China was too big for Europe to conquer, but too weak to join the Europeans in their expansion. Nevertheless, with the proliferation of global markets and the opening of commercial sea routes, China could recover after the collapse of the paper currency used during the Ming dynasty, which generated inflation and counterfeiting in the 1540s.

China accomplished this task by importing the Spanish "Carolus" dollar, minted using silver mined mainly in Mexico. These coins were a tremendous boost to the economy, which geared up to pay for the silver by exporting earthenware, porcelain, textiles, silk, tea, and more overseas. During this period, China accounted for 30–33% of global GDP, it imported about 30% of the world's silver production, and it was the world's largest economy. It is not surprising that the booming economy led to a corresponding increase in birth rates and population growth.

However, this pattern came to a stop in the early 19th century, when the gradual depletion of the silver mines in Mexico and Bolivia made extraction expensive. That caused political unrest that led the former Spanish colonies in Latin America to gain their independence. That led to a further decline in silver production, since smelting silver required mercury that was mined in Almaden, in Spain. The consequence was a dearth of silver in China that led to an economic crisis, internal rebellion, and the reversal of the silver trade direction. Now, silver started moving out of China in the form of strange boat-shaped ingots called "sycees." On top of that, the British government thought it could make good money by selling the Chinese opium cultivated in India and so squeezing out the remaining Chinese silver (incidentally, opium imports are normally considered the cause of the Chinese decline, but the problems had started much earlier)

The result was generalized chaos. In a famous incident that took place in 1839, Chinese official Lin Zexu seized large amounts of opium from foreign ships, then had it thrown into the sea. The British reacted by invading China. It was the start of the Opium Wars and of the "century of humiliation" in which China was repeatedly invaded and defeated by coalitions of foreign powers. The Qing government couldn't effectively counter the Western aggression, since it was engaged in quelling internal rebellions. In addition, China hadn't adopted the Western industrialization method, unlike what Japan did during the same period. So, the Chinese military forces were technologically inferior to the Western ones.

Perhaps it was the availability of low-cost human power that discouraged the use of production machinery, although bureaucracy and cultural factors also played a role. Whatever the case, China remained militarily and economically weak for a long time, and the Western attitude remained aggressive, not just in terms of periodic military attacks. Chinese immigrants in the US were subjected to all sorts of violence and intimidation. As an example of what was floating inside Westerners' minds, in 1909, Jack London, well known as the author of *White Fang,* wrote a short story titled *The Unparalleled Invasion* (1910). In it, he described the total extermination of the Chinese population by biological warfare. China is then colonized by the Western powers, opening the way to a "joyous epoch of splendid mechanical, intellectual, and art output."

The establishment of the People's Republic of China in 1949 marked the start of an economic and social renaissance and a new population boom. It was Chen Yun (1905-1995), head of Central Finance and Economic Commission, who redressed the monetary disaster by pegging the value of the Chinese Renminbi to real commodities that the government controlled: grain and cotton. Chen Yun's insight was revolutionary: he understood that currency is merely accounting for physical resource flows. By pegging the renminbi to state-controlled commodities rather than to scarce metals or foreign currencies, he broke China's centuries-old dependency on external monetary standards. This created the stable foundation for China's subsequent industrialization and population boom. The Korean War (1950 – 1953) saw China again fighting a coalition of

Western forces. The war ended in a draw, but it officially marked the end of the century of humiliation.

In the new phase of economic expansion, the Chinese population started a rapid growth, fueled by the introduction of new agricultural technologies. The government initially saw population growth as a good thing, and it enacted laws to limit birth control and subsidize large families. But already in the 1950s, worries about overpopulation were surfacing. The debate was affected by the "Great Famine" that struck China in 1960, during the "Great Leap Forward" designed to generate a rapid industrialization of the country. The reported number of victims from the famine has often been exaggerated, but today, estimations tend to converge to 15-20 million. A tragic event, in part due to droughts and bad weather, but in large part due to the misallocation of resources to industrialization, penalizing agriculture.

The Great Famine showed that the Chinese agricultural system was fragile. Nevertheless, the idea that China had an overpopulation problem was strongly resisted. During the Cultural Revolution (1966-1976), the natalist position was predominant. At the UN Conference on Population in Bucharest in 1970, a coalition of industrialized countries proposed that Third World Nations needed to implement birth control measures to avoid overpopulation. China sided with the Third World, forcefully condemning the proposal. The Chinese delegation maintained that world poverty was not caused by overpopulation, but by the imperialist policies of the rich world.

The Natalist position was often expressed in simplified terms as "more is better," and it was associated with the "Gang of Four," which dominated Chinese politics during the Cultural Revolution. These ideas fell out of favor after the death of Chairman Mao Zedong in 1976. With Deng Xiaoping at the helm, from 1978, the Chinese government completely reversed its position, rapidly pivoting to a set of policies and regulations designed to curb population growth.

How did the Chinese government arrive at this position? From what we can read in the 1981 book *China&Population, Problems&Prospects*, a collection of papers from Chinese scientists, it seems clear that it was the result of an extensive debate among scientists, administrators, and party members. The discussion was anchored on Marxist principles; for instance, noting how Engels himself had stated in 1891 that "a communist society should one day be compelled to regulate the production of human beings as it regulates the production of goods." Chairman Mao Zedong had also maintained in a 1957 speech that it was a task for the state to control the population in such a way as to optimize productivity. Apart from this ideological veneer, however, the debate in China was pragmatic. It was centered around a basic question: how should the population be managed to support the highest industrial output? Was it better to increase the number of workers or the productivity of each worker? The answer came from the studies carried out by the Chinese scientists working on demography.

From the beginning of its existence, the PRC government had encouraged scientific research with a view to competing with the West. In the 1970s, the Chinese school of demography went in parallel with the Western one in developing new statistical and quantitative methods of study. At that time, the main actors were Liu Zheng, director of the Research Institute of Demography of the Chinese People's University, and Song Jan, a scientific maverick who had started his career as a rocket scientist, later moving to demographic studies. Both Liu and Song were also prominent members of the Chinese Communist Party (CCP). Together with several other scientists, they framed the Chinese demographic future.

Liu and Song were part of different schools of thought: while Liu was a sociologist by formation, Song was a mathematician expert in control systems, a field sometimes termed "cybernetics." These two schools were partially in contrast with each other, and, in the West, cybernetics applied to social sciences was soon expelled and marginalized. In China, instead, conventional social sciences and social cybernetics seemed to be able to live together, probably because both claimed that their approaches adhered to the basic Marxist principles.

The 1981 book *China&Population* contains a long article by Liu Zheng that describes the goals of the Chinese demographic policy as part of the general planning of the economy. Liu strongly promoted the view that population growth should not become a burden for the progress of industrialization. In the same book, Song Jian presented similar ideas, but went into the details. First, Song evaluated the "optimal" Chinese population, estimated as 600-700 million. Then, Song sketched how this number could be reached in about one century after having overshot it to about 1.2 billion, and then carefully controlling the fertility rate of Chinese women in such a way as to avoid an excessive imbalance of the age distribution. Song published several more papers in English on his work in the international literature describing his mathematical model of population growth.[87] Song's book, titled *Population System Control*,[88] coauthored by Jingyuan Yu, provides more details.

It is difficult to say how strongly the Chinese government was influenced by the scientific arguments of Chinese demographers. Additional strategic considerations may have played an important role. From the documents of the time, it is clear that the Chinese leadership was seriously worried about the possibility of a new famine that could have destroyed the effort toward industrialization and military security of the country. In addition, the memory of the military failure of the Chinese against the Western invasions during the century of humiliation was still present in the minds of administrators. They may have reasoned that the reliance on cheap manpower during the Qing dynasty was one of the factors that prevented China from industrializing, and they didn't want to make the same mistake again. Whatever the case, after the downfall of the Gang of Four in 1976, a general agreement was obtained within the

Chinese Government that it was necessary to control population growth.

The Chinese birth-control action started in the mid-1970s. In an initial phase, no compulsory restrictions on family size were implemented; then, in 1980, the one-child policy was introduced. It was a set of regulations and administrative measures enforced by the National Population and Family Planning Commission. These regulations were never completely rigid, and they admitted several exceptions. In 1984, the policy was adjusted to allow a second child for families in rural areas if the first child was a daughter. Starting in 2006, in some provinces, couples who were both only children were allowed to have two children. In 2013, an exemption was introduced, allowing two children in families where one parent was an only child. In 2015, the Chinese government announced that all couples would be allowed to have two children, effectively ending the one-child policy. Today, the government is encouraging larger families.

But what was the effect of the one-child policy in China? Was it really effective, as it is commonly said to have been. Let's take a look at the data

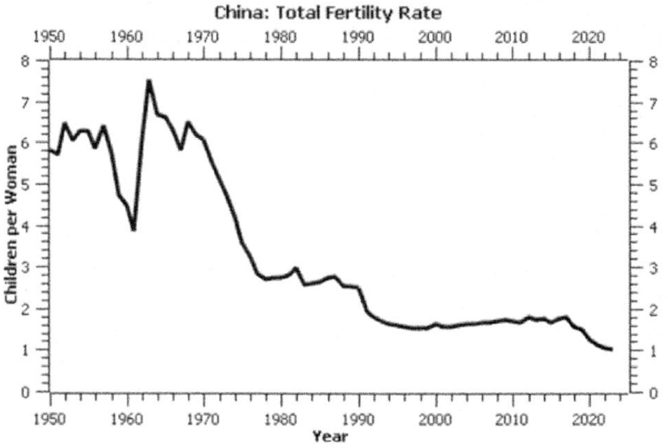

Figure 24. China's total fertility rate after 1950.

The strong oscillations of the fertility rate around 1960 are likely related to the famine of 1960 and the turmoil of the Great Leap Forward. People reacted to stressful conditions by reducing the number of births, and then by overcompensating for the previous decline. In any case, the early 1970s marked the beginning of a rapid decline in the fertility rate, which started well before the enactment of birth control policies. On the contrary, the TFR rose slightly after the one-child policy started, probably as a result of the large cohort of people born in the late 1960s entering their reproductive age. The elimination of the policy during the 21st century didn't generate a significant return to high fertility. Recently, China's total fertility rate (TFR) declined even further. Paradoxically, China reached the goal of one child per family only after the one-child policy was abolished.[88]

Further insight comes by comparing the Chinese fertility data to that of the four South-Asian "Tigers:" Singapore, Hong Kong, Taiwan, and South Korea. This comparison is significant because these countries are culturally and economically similar to China. Here is the relative graph.

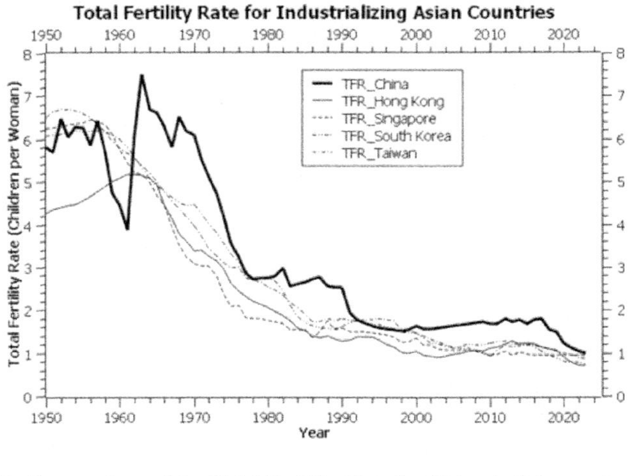

Figure 25 - Comparison of the Total Fertility Rate for China (solid curve) and some Southern Asian Countries

Remarkably, the fertility curves for all five countries are very similar, despite the fact that only China implemented a one-child policy. Today, all these countries have basically the same low fertility rate, well below the population replacement level. Does this mean that the one-child policy in China had no effect? Not necessarily, but, evidently, it was no more effective than the less drastic policy interventions to reduce natality enacted in the other four countries.

The initial Western attitude toward the Chinese birth control policy was one of interest and mild criticism. For instance, in a 1985 article,[89] Bongaart and Greenhalgh argue that the one-child policy was not necessary, but not against the necessity of reducing the Chinese fertility rate. Nevertheless, today, China is universally criticized in the West for the birth control policies enacted up to 2015. The criticism is based on four fundamental points.

- Government-enacted birth control is a mark of dictatorship and hence unacceptable on political and moral grounds.
- The one-child policy led to a large increase in the number of abortions.
- The one-child policy led to a strong imbalance in the sex ratio of newborn children.
- The one-child policy was counterproductive, leading to the current collapse of birthrates.

The first point of the list, unacceptable dictatorial policy, is a legitimate political opinion, but it is also used as an excuse for various exercises in China-bashing. As I discuss in a blog post,[90] articles on the one-child policy in the Western mainstream media liberally mix reality and fantasy, presenting a distorted view of the story in terms of a conspiracy somehow created by a coalition of Chinese Communists and Western evil forces. These articles appear to derive their criticism from an academic paper published in 2005 by Susan Greenhalgh.[91]

It would be out of place here to go into the details, but let me just say that Greenhalgh's thesis but let me just say that Greenhalgh's thesis effectively denies agency to Chinese demographers, suggesting they were either manipulated by Western ideas or incapable of independent scientific analysis—a patronizing view that ignores the sophisticated quantitative work being done by Chinese researchers throughout the 1970s.

In addition, Greenhalgh proposes a nefarious role of the Club of Rome, which she describes as a misguided organization of alarmists in her later book *Just one Child* (2008). However, nowhere does Song cite as an inspiration the *Limits to Growth* report that the Club of Rome sponsored in 1972, nor is there evidence of contacts between Song and the members of the Club. In any case, if Song read that report, he must have found it not very interesting. He was working on practical methods of population control, a subject that *The Limits to Growth* did not discuss. The best that Greenhalgh can do to support her interpretation is to state that "one exceptionally knowledgeable researcher maintained that Song had read and was directly influenced by the work of the Club of Rome" (note 17 at page 352 of her 2008 book). This source is said to be "confidential" and hence must remain anonymous. Why having read *The Limits to Growth* should be something confidential is surprising, to say the least. It is even more surprising if you consider that a Chinese translation was published in 1980 with a foreword by Premier Zhao Ziyang (then one of China's top leaders). But we are not here to discuss conspiracy theories, so let's move on.

The second item in the list of criticisms of China's policies is that the one-child policy caused an increase in the abortion rate in China. This is possible, although the data do not show a large effect.[92] The number of abortions did not significantly increase after the one-child law was enacted in 1980; they actually declined in the 1990s. The abortion rate picked up again after 2015, when the one-child law was abandoned, a trend hardly compatible with a direct effect of the one-child policy. In any case, modern contraceptive methods were widely available to Chinese couples from the mid-1960s onward. So, abortion was not the only possible way to comply with birth control policies. It may have been used nevertheless, because abortion is not normally seen as a moral issue in China, but as a practical birth control method.

The third item in the list is that the one-child policy caused, or increased, the gender imbalance in China, that is, the birth of more women than men. This is not a new problem for China, where a traditional preference for male heirs led to gender imbalance already during the Han dynasty (2nd century CE). From the Song dynasty onward, Confucian moralists repeatedly condemned female infanticide as a widespread practice. It seems that the highest imbalance was seen during the Qing dynasty (18th-19th century CE), when the male to female ratio was reported as about 110-150 to 100. Some provinces even reported values of the order of 200. These data can be found in the 1999 book by James Lee and Wang Feng, *One Quarter of Humanity*.

In ancient times, the gender imbalance was mainly the result of infanticide. In ancient China, "raising babies in the bucket" was a euphemism for drowning baby girls that seemed to be widely understood. In modern times, however, selective abortion can obtain the same result, since it is possible to determine the sex of a fetus before it is born by means of ultrasound examination. The Chinese census data indicate a slight increase in the male/female gender ratio after the introduction of the one-child policy in 1980, from the natural ratio of 105/100 to about 107/100. In the 1990s, when ultrasound techniques became widely available in China, the ratio rose to 114, and then to about 120 in the early 2000s. Since then, it has declined, being now around 110/100. For comparison, Taiwan now has a gender ratio of 107/100, despite never having implemented compulsory birth control policies. Today, induced abortion remains broadly legal in

China, but selective abortion is strictly forbidden. The gender imbalance is disappearing in urban areas, although it continues to exist in rural ones. We can conclude that the one-child policy may have contributed to an existing tendency to a higher male/female imbalance. Still, it was not the only factor involved, and perhaps not even the most important.

Finally, how about the accusation that the one-child policy was so effective that it is leading to the destruction of China by depopulation? (See, for instance, the 2021 article by Dominic Pino on National Review[93]). Honestly, it is impossible to take it seriously. It would mean that a policy that was abolished in 2015 is still affecting the behavior of Chinese families today, causing them to keep their fertility ratio at the levels that were required in 1980. A little unlikely, to say the least. After all, do the Chinese still use pots and pans as drums to chase away the invisible dragon that's eating the sun during eclipses?

In the end, what can we say about the Chinese experience in birth control? One conclusion is that the one-child policy was not an evil conspiracy to depopulate China. It was the result of well-founded concerns about the future of China, and it aimed at avoiding new famines and supporting an efficient industrial economy. It was devised and enacted on the basis of the best data and theoretical models available at the time. Unfortunately, it is well known that, in theory, practice and theory are the same thing, but, in practice, they are not. The Chinese planners of the 1970s were still reasoning according to the old idea that human birthrates are determined by an "urge to reproduce," and hence that a top-down intervention was critically necessary to keep the rates low. It was only in the early 1990s that it became clear that a widespread, sustained global fertility decline had begun, making birth control unnecessary as a way to reduce population growth. So, the one-child policy in China may be archived as a well-intentioned attempt, but ineffective and unnecessary.

Today, China is no longer facing the overpopulation problem it was facing half a century ago. Just like other industrialized countries, China's fertility rate is now well below the level needed to maintain the current population. In 2024, the Chinese population declined for the first time after the boom of the 1950s, and China is now losing about one million inhabitants per year. With a total population of about one and a half billion, such a loss cannot have significant effects on anything, but the decline may accelerate in the future.

A debate is presently ongoing in China on how to cope with the declining birthrates, just as it is happening in other Asian countries and in the West. Everywhere, governments are attempting to encourage births by financial and social incentives, but these measures don't seem to be having significant effects. That's not surprising because the social factors that lead to low fertility are deeply embedded in the structure of modern societies, including the Chinese one. In addition, China is affected by the same pollution factors that tend to reduce fertility everywhere by damaging the human reproductive system. Pollution is deeply embedded in the industrial production system, and the effort needed to abate it is enormous. No matter whether social or chemical factors are responsible for the decline (probably both), reversing the trend would require truly drastic actions. At present, it looks likely that the Chinese, just like other industrial economies, won't be able to do much more than mitigate the problems caused by an elderly population. Perhaps automation and robotics will provide the necessary workforce that was once provided by young human beings. But only the future will tell us how much governments will be able to use their influence to maintain a fertility rate able to stabilize the population.

Soviet Union: Economic Collapse

The collapse of the Soviet Union in 1991 was a classic example of a system that had overshot its resources. The Soviet Government had tried to do too many things at once: being a military superpower, managing a large empire, exploring space, and keeping pace with the richer and larger Western Bloc. It wasn't possible. The last straw was the invasion of Afghanistan in 1979, which led to ten years of indecisive war until the Soviet troops left the country in 1989. The campaign badly strained the financial resources of the state. Two years later, in 1991, the Soviet Union imploded. In a few months, it went from one of the two most powerful blocs on the planet to an ensemble of failed states, with an economy in shambles and no significant military power. Afghanistan is known to be an empire destroyer, and in this case, it confirmed its traditional role.

Miraculously, the collapse of the Soviet Union didn't involve significant bloodshed. The countries that formed the Union went their own way, new governments were established, and the disappearance of Communism as an official state ideology seemed to open the way to a new era of prosperity that could come by adopting the practices and economic structures of the Western world. But it wasn't to be so. The former Soviet Union states plunged into a dark world of poverty and economic decline. The largest country of the Union, Russia, didn't fare much better than the others. The government went nearly bankrupt, and its national currency, the ruble, was so badly devalued that for a few years it was replaced by the US dollar even in everyday transactions.

The economic collapse weighed heavily on everyone in the main former Soviet country, Russia. Although nobody starved, the quality of the diet of the average citizen worsened considerably. The loss of jobs led to depression, alcoholism, and drug dependence. Suicides spiked up. Both the fertility rate and the total population plummeted.

Fortunately, the bad times didn't last forever. Gradually, the economy recovered, the ruble became stable, and it returned to being used as a normal currency. The government restarted paying salaries to employees, the police cracked down on the criminal gangs, and the population began to grow again, although it didn't reach the peak level of 1990.

Figure 26 - Russian Fertility Rate. Data from Our World in Data.

Figure 27 - Russian population, data from Our World in Data.

Recently, the war against Ukraine has been affecting not only the Russian economy but also the fertility rate of Russian women, which has been going down. It is probably for this reason that in 2025,

Russia's Federal State Statistics Service *(Rosstat)* stopped publishing data that it collects on the country's demographic situation. Nevertheless, we know that the Russian TFR (total fertility rate) is now at a historical low of 1.4, well below the replacement rate of 2.1. And that's remarkable because 2024 was proclaimed to be the "Year of the Family" by the Russian authorities. Providing large families with preferential mortgages, tax refunds, free school meals, and discounts for public transport, among other benefits, was supposed to raise the TFR, but it didn't help. We may doubt that the planned "demographic task force" that the Russian government announced as being created in 2025 will do much better.

Overall, Russia is in the same quandary as most Western countries; no matter what the government does, the natality and the population continue to plummet. The collapse of the Soviet Union had much worse effects on Ukraine.

In Ukraine, the population decline was truly dramatic and took place in two phases. The first started immediately after the collapse of the Soviet Union. The second started with the conflict with Russia in the Donbass region. Note how, before the collapse, Ukraine was a relatively prosperous country, and the population was growing, reaching more than 52 million inhabitants. Today, it is around 38 million, the same as it was in 1950. Not all the missing Ukrainians are dead or never were born. Many emigrated to escape the war, and, in principle, they could return. But it is unlikely that the declining trend could be reversed.

Figure 28 Ukraine's total fertility rate. Data from Our World in Data.

Figure 29 – Ukraine's Population. Data from Our World in Data.

If we examine other countries of the former Soviet Union, we observe a trend. All the Asian republics, Uzbekistan, Turkmenistan, Kazakhstan, Azerbaijan, and Kyrgyzstan, show population growth. All the European republics, Belarus, Moldova, Latvia, Lithuania, and Estonia, show a decline, although not so sharp as Ukraine. In all these cases, the common factor that characterizes growth or decline trends is the availability of mineral resources. The Asian republics are rich in a wide variety of minerals. Kazakhstan has about 12% of the world's uranium reserves and is a strategic producer of enriched uranium as fuel for nuclear reactors. It has also copper, chromium, and more, plus oil and gas in the Caspian Sea region. Uzbekistan has gold and copper, plus oil and gas. Turkmenistan has some of the largest gas reserves in the world. Kyrgyzstan and Tajikistan have gold, silver, lead, zinc, and more. And, of course, the Asiatic part of Russia, Siberia, has vast mineral resources, including some of the world's largest oil and gas reserves.

Conversely, the former Soviet Republics in Europe have little in terms of mineral resources: Ukraine, in particular, has some coal and some iron, but neither is very valuable in the world's market. The fabled Ukrainian rare earth deposits, the object of much hubbub when Donald Trump started his second presidential term, simply do not exist as exploitable deposits, as you can read in the database of the United States Geological Survey (USGS). The story of these mythical "rare earths" that Trump was said to want as compensation for helping Ukraine fight Russia is a typical example of how fads appear in the Western mediasphere. For a while, they seem to be all-important, and then they just disappear in a puff of smoke.

Having mineral resources guarantees an income from exports that can keep an economic system alive even during dramatic times of political upheaval. A little prosperity helps people maintain their willingness to have children, and that's what's happening in the Asian republics and also in Russia. Except for Russia and Azerbaijan, they all have total fertility rates well above the replacement rate, and in one case, Uzbekistan, the rate is increasing – an exception in the world trends. Instead, the Former Soviet republics in Europe suffered badly from the collapse, and the declining economic situation discouraged families from having children.

The message that arrives from the former Soviet Union is that the economy is a fundamental factor in keeping society stable and avoiding a demographic collapse. Avoiding wars is also an excellent idea, and not just for this reason. That spells serious problems for Western Europe, whose economy is declining together with its fertility rate, with a serious risk of a Soviet-like general collapse. To say nothing about the rapidly growing warlike attitude of Western European Leaders. We will see where all that will lead us, but of one thing we can be sure: wars are not a good idea to improve a demographic crisis.

Ethiopia and Africa: The Other Side of Depopulation

Ethiopia is a fascinating country. It is in Africa, but it is unlike any other African country. On its high plateau, it has a long history largely independent from that of the rest of the world, almost like another planet. And it is one of the countries with the highest population growth rate in the world.

Throughout history, Ethiopia was a mix of cultures and races. In ancient times, it was a trading powerhouse, linking Rome, India, and Persia via the Red Sea Port of Adulis. In the 4th century CE, as the Roman Empire was collapsing, King Ezana of Ethiopia converted to Christianity, making Ethiopia one of the world's first Christian states, alongside Armenia and Rome. The Ethiopian Orthodox Tewahedo Church, rooted in this region, remains a cornerstone of national identity. Attempts by Muslim rulers to invade Ethiopia failed, just as the attempts by Europeans to convert Ethiopians to Catholicism.

In 1935, Ethiopia was the only country in Africa that was not under the control of a European power. At that point, Benito Mussolini, Italy's government leader, decided that the concept of *Lebensraum* (vital space) popular in Germany could be applied to Italy, too. Furthermore, the vital space that Italy needed (renamed "a place in the sun") could be found in Ethiopia. At that time, Italy had 42 million inhabitants occupying 300,000 square km. Ethiopia had 20 million inhabitants over an area of 1.1 million square km. So, there was a certain perverse logic in the idea of moving part of the Italian population to a much less densely populated region. But it was one of those cases in which a political ruler thinks he has a simple solution for a complex problem. It never works. The idea of invading Ethiopia generated an unmitigated disaster for all the actors involved.

From a military viewpoint, Italy managed to occupy Ethiopia in a relatively short time, even though the Ethiopian resistance was strong and had to be crushed by aerial bombing and poison gas. But the invasion nearly bankrupted Italy, and it was soon discovered that the poor Italian peasants had no intention of moving to an arid place where they would become even poorer. They correctly judged that they had more than enough sun in Italy. The Ethiopians, in their turn, suffered massive destruction and large human losses.

When WW2 arrived, the Italian troops in Ethiopia could not be resupplied from the mainland, and they were rapidly defeated by a coalition of British and Ethiopian *Arbegnoch* (patriot) fighters. One may wonder what would have been the final result of World War II if the more than 100,000 fully equipped Italian troops in Ethiopia had been available in the Mediterranean war theater. But this is not a book about historical fiction, so let's skip this subject.

The post-war period was difficult for Ethiopia: political upheavals and civil wars erupted. The 1984–1985 famine, claimed over a million lives. Nevertheless, Ethiopia's economy kept growing. At the same time, the population literally exploded, reaching 120 million in 2024. Note, however, that Ethiopia is nearly three times larger in terms of land area than its historical nemesis, Italy. So, in terms of population density, Italy (195 people per square km) is still more populated than Ethiopia. (114 people per square km.) The difference is that Italy's fertility rate is now below the replacement rate, while that of Ethiopia remains well above it.

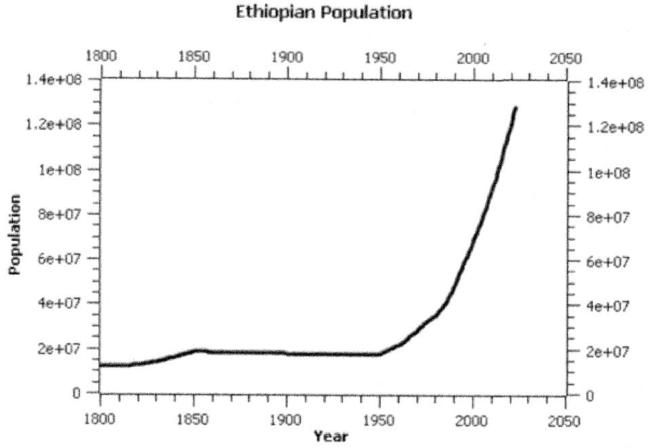

Figure 30 - Ethiopian Population. Data from the United Nations.

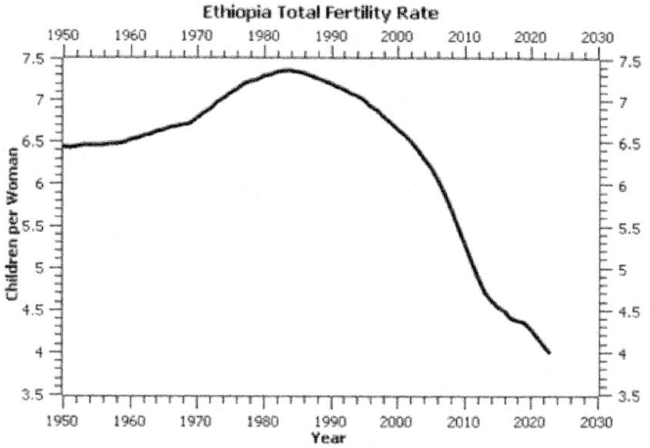

Figure 31- Ethiopia's Fertility rate. Data from Our World in Data.

In the 1980s, Ethiopian women had more than seven children each, a large country. Then, fertility declined rapidly, but it is still well above the replacement rate of 2.1 children per woman.

What about the future? With over four children per woman and a large cohort of young women able to reproduce, Ethiopia's population is destined to keep growing for some time, possibly throughout the 21st century, with the statistical projections showing that it could reach more than 300 million inhabitants in the early 22nd century, when it should start declining. These data look impressive, but, again, numbers have to be placed in context. With 300 million inhabitants, Ethiopia would still have a lower population density than Japan today.

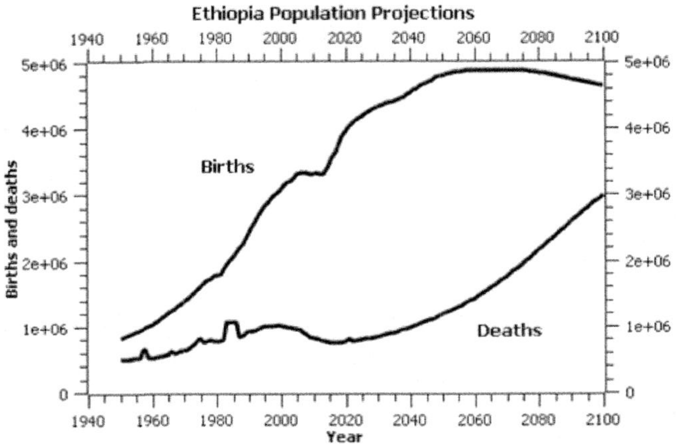

Figure 32- Ethiopia births and deaths, projected to 2100. UN data.

The population trends for Ethiopia parallel those for central Africa, but they are very different in the regions North of the Sahara Desert, where fertility is declining in the same way as in industrialized Western Countries. Overall, Africa's population trends are dominated by those of the central region, including Ethiopia

If nothing changes, Africa's population could reach about 40% of the world's population by the next century, up from the current value of 18%. Perhaps, Africa could reach a population of 4 billion people, and even more, according to some projections.

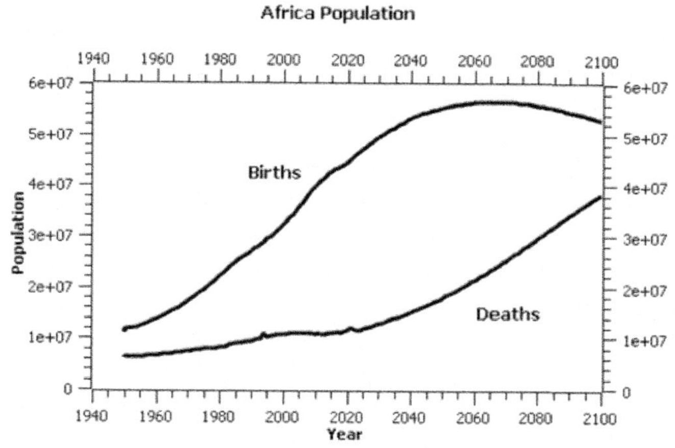

Figure 33 - Africa Birth and Death trends. UN data

Nevertheless, these long-term projections have to be taken with extreme caution. Whereas the North expanded in a period of abundant resources, largely provided by the South, now the South doesn't have an equivalent resource basin, and might well encounter a Malthusian barrier to reach population density levels comparable to those of the North. Eventually, the demographic trends in different countries will have to be compatible with each other, and we can only hope that it won't involve violence

The Future.

Technology can relieve the symptoms of a problem without affecting the underlying causes. Faith in technology as the ultimate solution to all problems can thus divert our attention from the most fundamental problem — the problem of growth in a finite system — and prevent us from taking effective action to solve it – Dennis Meadows

Figure 34 - A detail of the painting by the Italian painter Telemaco Signorini L'Alzaia (The Riverbank) 1894. It shows the plight of human beings used as energy sources. Wikimedia, Public Domain.

Taking a peek into the future of the human population, it is clear that we are approaching a "perfect storm" where several effects may coalesce to create a rapid population decline that we may call collapse. It is a thorny subject, loaded with political overtones of all sorts.

Predicting when and if the world population could crash down from the current levels turned out to be a honey trap for catastrophists, in which many people who should have known better fell, such as when Paul Ehrlich published his book *The Population Bomb* in 1968, predicting famines and assorted disasters in the short term. That book provided endless ammunition for the cornucopians to poke fun at the funny antics of catastrophists. Yes, but it is also true that famines and disasters did occur in the past, even though never at the planetary scale and not during the past few decades. So, without falling into the trap of predicting dates, what can we say about the population trends in a not-too-remote future? Let's play the scenario game.

Food Production: Malthus' Return?

Food is the most obvious element that determines the population level, and its availability will be a crucial factor in the future of humankind in the coming decades. The limit to food production on a given extent of surface varies depending on several factors, but there is an upper limit given by the flow of solar light that powers photosynthesis. The actual limit is usually much smaller, given by other factors such as the availability of nutrients, water, and the range of temperatures in which plants can live and prosper. It also depends on the efficiency with which the products of the land or the sea can be harvested, processed, and stored. We can call this production level the "Malthusian Limit." It is not an absolute limit, but there is no doubt that it exists.

Ancient civilizations operated close to their Malthusian limit, as shown by recurrent famines. It was on the basis of this observation that Malthus proposed his model. But how close are we to the limit in our modern times? The data about global famines show a declining trend that started in the mid-20th century. Evidently, during the past few decades, the world's agriculture could match the demand for food worldwide. Some people took such data as evidence that famines have disappeared forever. One was Julian Simon with his *The Ultimate Resource* (1996). A more recent work by Steven Pinker, *The Better Angels of Our Nature* (2011), deals mainly with the disappearance of violence and conflicts, but also proposes that famines have disappeared for good.

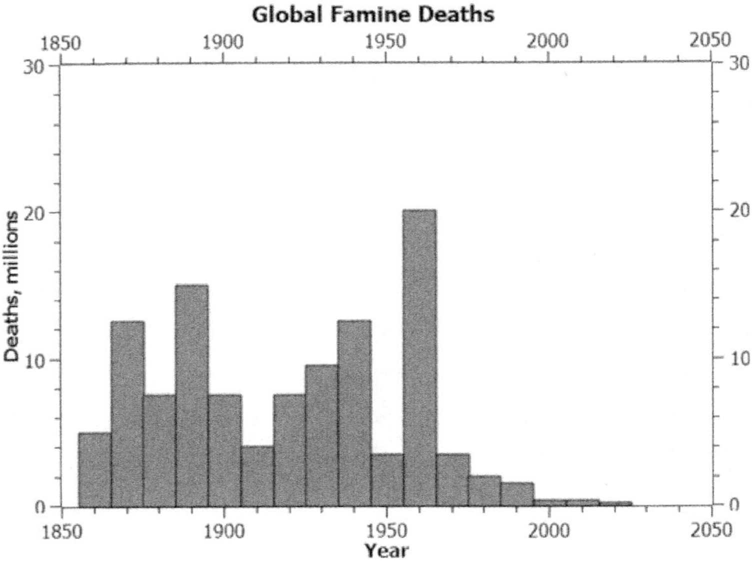

Figure 35 – Famine victims worldwide

At this point, you could argue that Malthus was dead wrong. But that would be an excess of optimism. Necessarily, for a famine to occur, it must be preceded by a period of abundance that leads the population to increase beyond its limits. It is reported that, in the 1840s, potato harvests were so abundant in Ireland that some had to be thrown away into ditches. You might have taken that as proof that Ireland would never suffer from famines again, but, alas, that wasn't the case. The great famine of 1846 was just around the corner.

The fact that we haven't seen global famines ongoing during the past 75 years doesn't mean we won't see them again. A lull of less than a century in a history that spans several thousands of years is not enough to state that the worst devils of our nature aren't waiting in the shadows to hit us again. Even if the risk of large-scale famines does not seem to be immediate, we do have a problem of food security, with many people, even in the rich world, who aren't sure that they will have enough to eat every day. According to FAO, today, about 10% of the world's population suffers from severe food insecurity, and almost as many are undernourished. Many of the remaining fraction are not appropriately nourished because obesity and type II diabetes are rampant diseases. If the trend continues, half of the world's population is going to suffer from one or another form of diabetes, including the dreary "type III" form, which is the way some people call dementia. These trends are increasing, according to FAO.

Several possible conditions could lead to a contraction in the availability of food for a large fraction of the world's population. The main factor is, of course, the economic system's capability to keep producing food at the current levels. In this section, I will concentrate on agriculture and the associated livestock raising, leaving the other main food supplies, fishing and aquaculture, aside since they are minor sources in comparison, or largely depend on agriculture for their existence.

Today, agriculture has grown to a major planetary force. It is highly mechanized with the use of tractors, harvesters, and other machinery, significantly reducing manual labor and increasing the scale of farming operations. Industrial methods have led to specialization, where farms focus on a single crop or animal species to optimize efficiency. High-input agriculture is where large amounts of resources like water, fertilizers, and pesticides are used to maximize yield per unit area. The use of Genetically Modified Organisms (GMOs) is expanding, with genetically engineered crops leading to pest resistance, herbicide tolerance, and improved nutritional content. The newest trend is "Precision Agriculture," where farmers use technologies like GPS, sensors, drones, satellite imagery, and the like to optimize parameters such as the application of water, fertilizers, and pesticides. Theoretically, this approach reduces resource use and the environmental impact.

These improvements are all part of a series of changes promoted to make it possible to feed a growing number of people. But agriculture is a complex system that involves a variety of technologies that interact with each other, not always with good results. Consider GMOs: sometimes, they are lionized as proof of human ingenuity that reaches stellar levels. GMOs can indeed increase the amount of calories produced per unit surface in industrialized agricultural systems, but the result is not necessarily positive. The use of GMOs generates a cascade of effects that reverberate in the whole system, typically transforming a resilient rural production system into a fragile, heavily industrialized system. To say nothing about the fact that these technologies often worsen the food's nutritional quality.

To exploit the increased yield of GMOs, you need fertilizers, pesticides, irrigation, agricultural machinery, storage facilities, and more. Even if you can manage to upgrade all that, very often, the result is to make the system more vulnerable to external shocks, for example, increases in the cost of supplies, such as fuels and fertilizers. An economic crash or an increase in the cost of fossil fuels from which pesticides and fertilizers are created might make GMOs as obsolete as steam locomotives.

There are other examples of deeply flawed strategies in food production. One is the idea that we can increase supply by eliminating food waste and arriving at or near a "zero waste" condition. This idea is often presented together with images showing how a large amount of food, nowadays, is thrown away into waste bins, especially in Western countries. Sometimes, though, a little waste may not be a bad thing, and it is well known that perfectly optimized systems are also perfectly fragile.

As an example, years ago, the student cafeteria of the University of Florence had a conveyor belt that went through the hall and carried the trays with the lunch leftovers back to the kitchen. The trays often contained plenty of edible food, and some of the local poor had taken the habit of sitting near the end of the conveyor belt to catch the food before it disappeared into the kitchen. They were a quiet presence that bothered nobody, and some good-natured students would purposefully leave some intact food for them in their trays. Unfortunately, the idea that the poor would eat the students' leftovers was considered unseemly and somewhat subversive by the university managers. The cafeteria was rearranged in such a way that the students themselves would bring back their trays to the entrance of the kitchen in the same way as is commonly done in fast-food restaurants. Then, the edible leftovers were thrown away. A big improvement, don't you think?

This story highlights how our society is poorly organized to cycle its resources, and still operates according to obsolete dreams of abundance. Much food is wasted in restaurants, but collecting it for the poor is not often done. Sometimes it is forbidden by law, although sometimes it is mandated, such as in France. It is a complex story that varies from country to country, involving different rules, laws, and cultural factors. An interesting story is that of the Canadian "binners," the people who collect food from waste bins. The practice is forbidden by law, and it is reported that sometimes the police would stand on guard in front of waste bins or run after binners. Surely not a way to optimize the use of societal resources.

We see the same story for traditional gleaning; once a common feature of peasant societies, today it is sometimes tolerated, but more often forbidden. In the 1930s, in the Soviet Union, gleaning was declared a criminal activity that could be punished with death. Fortunately, it is reported that the law was rarely applied in full. Nevertheless, reusing food waste is sometimes seen as degrading for the poor and offensive for everybody, as it goes against the idea that there exist technological solutions for everything. In practice, eliminating food waste would not just be impossible, but it would involve high costs and would make the system more fragile and vulnerable to shocks.

That doesn't mean that waste is unavoidable. Natural ecosystems are zero-waste systems; they recycle everything. But they are organized in a different way from the human food production system. An ecosystem is an ensemble of subsystems that work on the principle that "waste is food." That is, what is waste for one subsystem is food for another. But we humans are far from being able to do the same. The industrial system would have to be rebuilt from scratch, imitating the way natural systems work. But, at present, it is not done at a significant scale anywhere in the world.

Another apparent solution to food scarcity is expressed as, "If we all went vegetarian, there would be plenty of food for everyone." The production of animal proteins is indeed a wasteful process that uses agricultural products as feedstock for animals. If the areas presently used for livestock were turned to agriculture, more food would be available, at least in terms of calories. But switching to vegetarian food has plenty of problems, the main one being the difficulty of providing sufficient protein. Furthermore, "green" food is bulky and hence expensive to transport. It also spoils easily, and so you need refrigeration, and so on.

Similar concepts are shown by the "let's go local" strategy, that is, eat only food that's produced near where you live. That eliminates the cost of transportation, but involves huge problems, too. How do you deal with the unavoidable fluctuations in local production? Once upon a time, these fluctuations were the cause of periodic famines, often accepted as a fact of life. Going back to that is not exactly a way to solve the food supply problem.

How about the newest poster child of agricultural technologies, precision fermentation? It is a biotech process that uses microorganisms, like bacteria, yeast, or fungi, to produce high-value chemical compounds. It is a little miracle of genetic engineering that allows people to hack the DNA of bacteria to insert the sequence that generates the protein they want. Each bacterium then becomes a tiny factory that churns out whatever the DNA sequence is assembling.

The modified microbes go into a bioreactor: a water-filled tank with nutrients (usually sugars or starches), and the right temperature and acidity. It's like a brewery vat, but instead of beer, the vats grow custom molecules that can be harvested as food. You may get, for instance, an artificial burger or artificial cheese. You can also indulge in spectacular claims, such as creating mammoth meat, a food that no human has tasted for probably a few thousand years. Actually, the "mammoth meatball" story was just an advertising stunt. We don't have the DNA sequence of extinct mammoths, so it was impossible to recreate real mammoth meat.

Impressive as these technologies are, what can they do in addition to giving us the illusion of being paleolithic hunters? There are reasons why precision fermentation could be a game-changer in several fields of food production. Suppose you want to create meat proteins. You need to raise cattle, pigs, or chickens, and then kill them. Hopefully, it is done "humanely," but it is still killing living beings. Moving the production of proteins to fermentation plants, you could re-naturalize huge swaths of land with great advantages for the health of the planet. It is also fast: you can produce a lot in a short time, without worrying about changing seasons and the natural cycles of animals.

On the other hand, fermentation plants function with anaerobic organisms that don't perform photosynthesis. So, what they can do is to transform sugar into more sophisticated food. But they can't do what plants do: produce sugar out of atmospheric CO_2. You still need plantations to do that. Can we remove the dependence of the bioreactors on cultivated sugar? Yes, in principle, you could utilize a less valuable vegetable product, such as corn stalks. With appropriate genetic programming, the bacteria could break down hard plant biomass into glucose and then use it to manufacture food. Some bacteria (e.g., gluconobacter) already make gluconic acid (a glucose derivative) commercially via fermentation. Marketing these products, we would become able to eat wood as termites do. A major metabolic change for humans.

With some creativity, we could also direct bacteria to transform electrical energy into food. Renewable energy plants produce plenty of electric energy, and in periods of low demand, this energy comes at zero cost. There doesn't seem to be a way to power the natural photosynthesis machinery of plants using electrons. Still, electrons can be turned into photons with good efficiency using, for instance, LED lights. Using electric energy directly produced by renewable plants, or stored in batteries, it would be possible to create farms completely independent from solar light. Then, it may be possible to hack the DNA of blue-green algae and make them produce food for humans in bioreactors, just as is done today using biofermentation. It is not clear whether that would be more efficient than using natural plants, but it could have significant advantages in comparison to traditional agricultural processes.

All is well and beautiful with these ideas, but one must also take into account that humans need not just sugar and proteins. They need all the micronutrients that plants obtain from the living soil. They need nitrogen, phosphorus, potassium, magnesium, sodium, and many other minerals in small amounts. They also need vitamins. And here things get complicated: we can instruct bacteria to create the basic proteins of meat or milk, but natural food contains compounds in microscopic amounts whose role in nutrition is probably important, but not completely understood. One thing is to eat natural meat that once was part of a living animal; another is to eat meat in the form of proteins created by bacteria. They may look the same, and even taste the same. But in terms of trace nutrients, they won't be the same. So, can people thrive on a diet of bacteria-made food? We don't know, and it will be an interesting experiment to do.Another problem with food production is that the modern food industry can be seen as a way to transform fossil fuels into food with some help from sunlight. Fuels are needed in the fields as fertilizers, pesticides, and fuel for agricultural machinery. Once the food is produced, it needs fossil fuels for transportation, processing, refrigeration, packaging, and storage. Agriculture directly uses about 3-5% of global fossil fuels. The broader food industry, including transport, processing, and packaging, accounts for roughly 15-20%. Numbers vary by region and efficiency, but these are the numbers. And you see what the problem is: fossil fuels are not and cannot be eternal, and they are also the main source of carbon dioxide, generating global warming. Right now, if you take away the fuel, billions of people starve.

Figure 36 - The author, Ugo Bardi (right), with Paolo Pasquini (center) and Toufic El Asmar (left), together with the RAMSES electric tractor. Photo by the author

Eventually, we'll have to learn how to produce food and supply it to consumers without using fossil fuels – it is not easy, but it is possible. My coworkers and I discussed that in a paper aptly titled *Turning Electricity into Food*[94] We even built an electric agricultural truck/tractor to demonstrate how it could be done.

Despite the efforts to push agriculture to a lower reliance on fossil fuels, the problem is rarely taken into account in agricultural planning. When fuel prices rise, the immediate reaction of farmers and government officials alike is to provide subsidies to reduce the impact. But that means simply shifting resources from one sector of the economy to another. It doesn't change the fact that fossil fuels will not last forever and that they are already doing enormous damage to the ecosystem with their emissions of pollutants.

In a longer time frame (but maybe not so long), we have to consider how climate change is going to affect agriculture, and how it is already affecting it. Climate change is disrupting weather patterns and creating havoc in traditional cultivations almost everywhere in the world. Droughts, in particular, are becoming harsher and more frequent in many regions of the world. The problem can be eased by switching to different crops, and it is at least in part compensated by the warming effect that makes it possible to produce cereals in northern regions, where it was too cold to do so in the past. On the whole, extreme effects of climate change are said to be unlikely to happen for at least a few decades, although we cannot exclude an acceleration of the change, which would make large parts of the world's regions too hot and dry to be cultivated.

In general, plants can't function at temperatures below 0 °C, and their vegetative rate drops rapidly above 35-40 °C. Some plants can survive up to ca. 50°C, but most will dry out and die over 40°C. This means that global warming may push temperatures in tropical regions to a level at which cultivating staple crops like maize, rice, and wheat will be difficult, if not impossible. Temperate regions may initially benefit from longer growing seasons and thawing permafrost, but beyond 1.5-2°C warming, heat waves and drought will reverse these gains. Adaptation strategies like heat-tolerant crops (e.g., drought-resistant maize), precision irrigation, and sustainable aquaculture can mitigate losses, but only within some limits. In general, the future of agricultural food production does not look good unless some radically new technologies can come to the rescue, or climate change can be contained.

Food Supply: Who's Going to Starve First?

Most of the discussion on providing food for humankind focus on the idea that we need to increase agricultural yields to increase food production. But that's just one of the elements of a problem embedded in a complex and delicate production system. Basically, the question is not just food production, but *food supply*. Two very different concepts.

What moves food from fields to plates is not the benevolence of the farmer (as Adam Smith would have said), but the capability of consumers to pay for it. This implies a working global economic system that allows people to perform some activity to earn money to buy food. The case of the Irish famine of the 19th century teaches us what the problem can be. There was food available in the world market, but the Irish didn't have the money to buy it. Hence, they starved to death. Both governments and charities can provide the money needed to help the poor, but they may not have enough or be unwilling to provide it.

The modern food processing activity is a long chain of linked activities: harvesting, processing, storing, refrigerating, transporting, packaging, marketing, and more. If just one of the steps fails, food does not arrive on plates, and people starve. A possible cause of the breakdown is the fragmentation of the global economic system into separate regions that engage in making war (commercial or kinetic) with each other, which is already happening. So, despite all the technological progress in agriculture, the situation is fragile and potentially heading for a disastrous outcome. We'll have to see how it will be possible to manage it, if it will be possible at all. Suppose, for instance, that the global economic system falters and collapses in the midst of wars and financial conflicts. Every country will have to rely on its national resources for food production. And, of course, many countries have a production deficit, and their people can have enough food only by importing it. So, who's going to starve first?

The question is not easy to answer because most of the data reported in the various sites dealing with food import/export use monetary units, and do not distinguish between the different nutritional content of agricultural products. So, you could think that Italy has an excess of food production because it exports wine and other expensive agricultural products, and hence, the financial balance is positive in terms of food imports and exports. But, of course, not even Marie Antoinette would ever say that if the people don't have bread, they can drink Chianti wine.

For a more realistic assessment of the food movements across borders, we can use the FAO trade statistics data to convert exports in weight and monetary units into calories. Even that is a partial index because people don't just eat calories — they need good food rich in nutrients. But that will do as a first approximation. I calculated the result for 13 countries that seemed to me representative of extreme cases, or of large and important regions. The results are in the table below, based on commodity volumes and average caloric densities (e.g., cereals at ~3,300 kcal/kg, soybeans at ~350 kcal/kg). The data excludes non-foods such as wine and other alcoholic beverages. These are rough estimates, not official figures.

Country	Pop (M, 2021)	Food Exp (Tcal/y)	Food Imp (Tcal/y)	Exp-Imp (Tcal/y)	Cal def/surp per cap (kcal/d)
Japan	126.0	1.5	28.0	-26.5	-576
United Kingdom	67.0	3.0	14.0	-11.0	-450
Saudi Arabia	35.0	0.5	5.5	-5.0	-391
Italy	59.0	4.2	10.5	-6.3	-293
EU (Total)	447.0	85.0	120.0	-35.0	-215
Kenya	54.0	0.8	4.2	-3.4	-172
Mexico	129.0	4.5	11.0	-6.5	-138
China	1,425.0	20.0	65.0	-45.0	-87
Ethiopia	117.0	1.2	3.8	-2.6	-61
India	1,391.0	15.0	18.0	-3.0	-6
Russia	146.0	18.0	12.0	+6.0	+113
United States	332.0	45.0	25.0	+20.0	+165
Brazil	213.0	22.0	8.0	+14.0	+180

Table 5. Estimated annual food calorie trade (trillions of kcal) for 13 selected countries, showing imports, exports, and import-to-export ratios for a number of countries. Based on FAO trade statistics. Data spans 2020-2023, sorted by increasing ratio, starting from net exporters (e.g., Brazil, ratio 0.05-0.13) to net importers.

The second and third columns list the food imports and exports in trillions of calories (or tera-calories). The subsequent column is the difference between imports and exports. A negative value means that a country has a deficit in calorie production. The last column lists the deficit/surplus in terms of calories missing or in excess per person. So, the countries at the top of the table are those which are most dependent on imports and whose citizens would be the first to starve if worse came to worst. China, Saudi Arabia, and Japan are in a bad situation. Instead, Brazil, the USA, and Russia are doing very well. Surprisingly, even India does reasonably well, despite its large population and its fame as a place where people often experience famines. That's the miracle of the green revolution and its fertilizers, pesticides, GMOs, and the like.

The table above gives only a partial view of the situation. As we saw in a previous chapter, agriculture needs energy to produce, and this energy nowadays comes mainly from fossil fuels. The countries that can produce large amounts of food today wouldn't be able to do that in a situation of a lack of energy supply, currently mainly in the form of fossil energy. So, I took into account this factor to evaluate national food security risks amid potential global supply disruptions. Let me summarize how I did it.

First, I started from the same data from the Food and Agriculture Organization (FAO) used to create table 5. Fossil fuel dependency (FFD) was estimated directly from 2022 energy import dependency ratios sourced from the World Bank and International Energy Agency (IEA), expressed as percentages of total primary energy consumption; for net energy-exporting nations (e.g., Brazil, Russia, United States), FFD was capped at 0% to reflect negligible import risk. The adjusted net balance, accounting for vulnerability in domestic production to energy shortages, was calculated as Adj Net = Net × (1 - FFD/100). This is an ad hoc hypothesis that serves to establish a reasonable weight of the use of fossil fuels in agriculture. Finally, per capita risk was derived by dividing Adj Net by 2022 population estimates (from United Nations data) and scaling to daily kcal/person: Adj Net Per Cap = (Adj Net × 10^{12} kcal/Tcal / 365 days) / population.

In table 6, countries were ranked by Adjusted Net calories imports (most negative first) to prioritize those facing the greatest aggregate caloric shortfalls. The table also shows the relative risk per person, taking into account the population of each country. (data from 2022 from FAO and World Bank)

Country	Imp Tcal/y	Exp Tcal/y	Net Tcal/y	FFD %	Adj-Net Tcal/y	Adj-Net/cap kcal/d
Saudi Arabia	-25	2	-23	0	-23	-1749
EU (Total)	-800	400	-400	62.5	-150	-919
China	-600	100	-500	20	-400	-776
Mexico	-60	20	-40	16	-33.6	-730
UK	-30	10	-20	34	-13.2	-539
Kenya	-15	5	-10	22	-7.8	-395
Japan	-120	10	-110	87	-14.3	-313
Italy	-40	15	-25	77	-5.75	-267
India	-250	150	-100	35	-65	-126
Ethiopi a	-10	5	-5	11	-4.45	-102
United States	-80	180	100	0	100	822
Russia	-50	250	200	0	200	3803
Brazil	-120	530	410	0	410	5221

Table 6 – Food imports and exports taking into account the production of fossil fuels

The new table shows several interesting elements. Now, the countries at risk in terms of food resilience are those that need to import fossil fuels. It is because agriculture needs these fuels. Do not pay too much attention to the absolute numbers; the influence of fossil fuels was estimated only qualitatively.

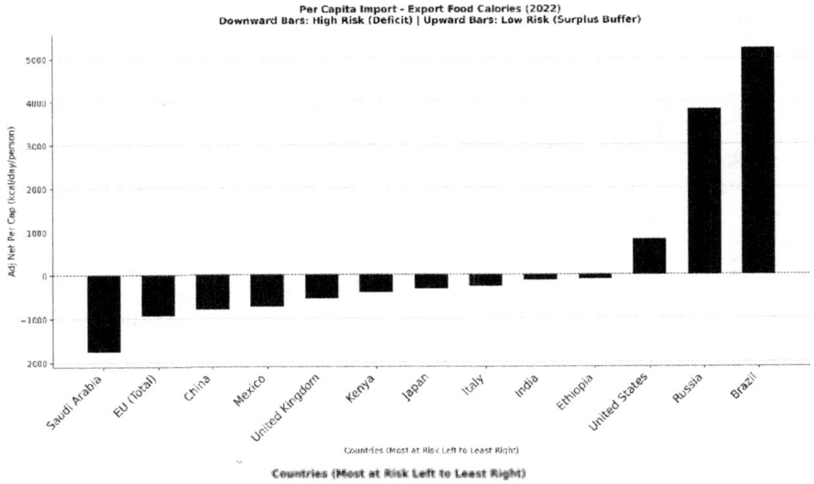

Figure 37 – Net food production for selected countries, taking into account the production of fossil fuels. Upward bars: low risk of starvation. Downward bars: high risk.

But the table tells us that Brazil and Russia have a positive adjusted net import-export ratio when you take into account fossil fuel production, and the same is true for the US. All of them have abundant fossil fuel resources and plenty of available land. The Netherlands is the only European country with a positive balance because of the availability of offshore natural gas. All the other countries have negative net import-export calorie values. Saudi Arabia has a highly negative net value despite being a fossil fuel exporter because it produces very little food. Note that a person needs 2,000 to 2,500 calories per day, even though, in practice, people in industrialized countries are accustomed to more than that. In any case, the results shown in the table are better visualized in a bar chart. The chart indicates that in the case of a major disruption of the global commercial system, most EU countries could lose a significant fraction of their calorie supply per person. In other words, people would starve. China faces the same problem, and this is probably the reason why it is pushing so hard on renewable energy to guarantee the energy supply needed for agriculture.

Of course, these calculations have to be taken with a lot of caution. One problem is that the table does not take into account climate change. Increasing temperatures and droughts are expected to affect countries closer to the equator, so, again, they should favor Russia. At the same time, they could badly damage India, which depends on its yearly monsoon for the water needed for cultivation. Even Mediterranean countries are expected to suffer from droughts according to the models used to calculate the effects of global warming.

These are the negative factors that could affect food supply in the future. But, of course, things never stand still, and, in the future, depletion might considerably change the values and bring some producers to the negative side of the food supply balance. Consider also that agriculture adapts, changes, and improves over time, together with climate, social, and political changes. So, the negative effect of climate change could be countered by adaptation, while new agricultural technologies could benefit from the low-cost renewable energy that can be produced nowadays using photovoltaic and wind plants. Nevertheless, the risk of a Malthusian catastrophe in a non-remote future cannot be completely discounted. As usual, the future always surprises us.

The End of Economic Growth

How is the economy going to react to a population decline? A case in point is Japan, whose population peaked at 128.1 million in

2008 and has been sliding down ever since, being at around 123 million in 2024. Fertility in Japan is stuck at 1.3 births per woman, well below the 2.1 replacement rate. According to data from the Japanese Ministry of Internal Affairs and Communications, in 2024 alone, the loss was about 899,000 inhabitants, close to 1% of the total. It corresponds to a drop of about 4.3% in 15 years, with projections of a 16.3% decline by 2050.

However, Japan remains a prosperous country. It is no longer the engine of growth it was a few decades ago, but it still has some of the world's highest standards of living and highest life expectancies (85 years). So, Japan shows that it is possible to keep an industrialized society alive and functional while losing about 1% of the population every year and, at the same time, moving the average age of the residents toward geriatric levels. Evidently, a population decline does not necessarily lead to catastrophic situations that could look like a Mad Max movie. Some studies based on economic models indicate that an aging population generates a decline in GDP growth, but not necessarily an absolute decline.[95]

But what about faster rates of decline? At present, we have no examples among modern industrialized countries of population declines so rapid and abrupt as to have caused the complete collapse of a country's economic system. The worst example is that of Ukraine, which, as we saw in a previous chapter, lost about 4% of its population per year after the collapse of the Soviet Union, and even more during the current conflict. Ukraine was able to survive as an organized state, but only with the support of a massive inflow of capital from Western States.

There are worse cases, even though not among industrialized countries. There are plenty of "failed states" in the world; states that theoretically exist, but where the government does not rule or control most of its territory. However, none of these failed states appears to have been associated to population decline. For instance, in the 1960s, Syria built a relatively modern state by exploiting its crude oil resources. It even arrived at a point where its government could dream of a "Greater Syria," challenging in military terms other local powers, including Israel. But, in the 1990s, with the decline of oil production caused by depletion, the Syrian economy declined, too, and the government had to stop dreaming about territorial expansion. It was still possible to keep the state functioning by turning Syria into a world manufacturing center of Captagon, a powerful amphetamine, illegal in most countries. Even that activity collapsed in 2024, possibly because of internal rivalries or sabotage by external powers. In 2025, the government was unable to pay its soldiers and the Syrian army collapsed in a few days when the country was invaded by a less-than-impressive force formed by a coalition of Sunni Islamist organizations. Now, Syria is fragmented into several semi-independent regions, in part controlled by foreign states. Nevertheless, Syria's population continues to increase, having reached nearly 26 million people in 2025.

From these examples, we see how difficult it is to predict the effects of the ongoing demographic decline when it enters high gear and rapidly drives down the population. Economists are obsessed with growth and see as a disaster the possibility of even a moderate reduction in the growth rate, for instance, from 2% to 1% annually in nations like Germany or Italy by 2040. But, most likely, we are not facing just a decline in the GDP growth rate. We are facing simply a decline that some people politely call "negative growth." That may already be happening, but it is masked by the adjustments made to the GDP, for instance, by including "hedonic factors," designed to make it always grow. With these arbitrary corrections, the GDP may no longer faithfully represent the state of a country's economy.

The problem is that a decreasing population reduces the labor force and also the demand for goods and services. Industries like housing and retail suffer first; fewer people need fewer homes. Deflationary pressures may emerge as businesses cut prices to attract scarce consumers, further squeezing profits. Unexpectedly, per-capita GDP could rise if productivity gains outpace population loss, as seen in some Scandinavian countries with high automation. Yet, without policy shifts, such as higher immigration, incentives for births, or extended working ages, most developed economies face a significant structural drag on growth.

In the future, labor dynamics may shift toward automation and outsourcing. McKinsey estimates that automation could raise global productivity by 0.8-1.4% annually through the 2030s,[96] and this is a conservative estimate. It could be much more. For Western countries adopting robotics and AI is critical to sustaining GDP with fewer workers. However, workers displaced by automation are also a problem. Most tasks that people normally perform in their jobs involve nothing more than moving things from one place to another. These things may be physical entities (say, you are a mailman or a truck driver), or numbers/strings of letters (say, you are a clerk, or a government bureaucrat). Or, maybe, your job is to try to move a chunk of lead from the inside of a gun barrel into someone else's body, but it is basically the same idea. Human brains have a certain capability to evaluate the position of things in space and to act on them. And that's what they are used for. But, recently, automated systems have become capable of performing these tasks better and faster, including the skill involved with putting chunks of lead inside people's bodies.

So, what kind of jobs can be found for people whose skills have become useless? Apart from removing them from the job market by means of the robotic skill of moving chunks of lead around, the very idea that everyone must have a job of some kind may be challenged. Yet, the current political views emphasize concepts that may be totally obsolete in a world where robots and AI will take over tasks that once only humans could perform. Politicians seem to promise us a world in which men will be the breadwinners, going to work every day, while women will be sweet wives waiting for them at home, taking care of the kids, and preparing good meals for the whole family. That world, alas, looks now as remote as the science fiction magazine covers of the 1950s, including the robot-butler vacuuming the carpet and carrying the Margaritas around.

In the future, pensions will also face a crisis as population decline accelerates. With fewer young workers supporting more retirees, dependency ratios worsen. In the EU, the old-age dependency ratio (retirees per working-age person) is projected to rise from 32% in 2020 to 50% by 2050. Pay-as-you-go pension systems, common in Western nations, are heavily strained under this imbalance. For instance, Italy's pension spending already exceeds 15% of the GDP, and a shrinking workforce could push that to 20% by 2040, risking insolvency without reforms like higher taxes or reduced benefits. Shifts with longer work hours are also considered as a response. [96] A smaller labor pool forces employers to compete for talent, potentially raising wages but also increasing per-worker hours. In Germany, where the working-age population could drop by 20% by 2050, firms already report labor shortages. Some companies push for 42-hour workweeks over the standard 38. As you may imagine, not every worker is happy about that. Alternatively, part-time work may rise for older workers, as governments incentivize delayed retirement. However, cultural resistance to these ideas is strong.

We must accept that in a stable or slowly declining society, we will have a large fraction of elderly people. We need to help them to remain in good health, and active and productive citizens. That will maintain the society's economy alive, and will also be good for the elderly if we stop considering them a burden. During the long history of humankind, grandparents have been a critical asset for families to help parents raise and educate their children. Nowadays, we tend to lock grandparents into nursing homes, where their only perspective is to watch TV while sliding into a vegetative state until they die. It is a lost opportunity for the young generations to learn from their elders, just as it is a terrible financial burden for families. Remedying this disaster is a huge challenge for society, but we must start thinking about it. MORE GENERATIONS LIVING TOGETHER ARE NOT ALWAYS HAPPIER. MANY ELDERLY PEOPLE LIVE IN THEIR OWN HOMES.

Figure 38 - An Italian Family in 2021. From right to left, Liliana (100), Grazia (68), Donata (33) and Aurora (1). Grazia is the wife of the author of this book. Photo by the author.

Plenty of horror stories can be found on the Web about how the elderly are mistreated in these homes. This phenomenon is well known in the scientific literature, and it even has an acronym, EA, for "elder abuse," which can go from simple neglect or verbal abuse to much more serious forms involving emotional, financial, and even sexual abuse. Some recent reports are truly chilling, for instance, a recent survey carried out in Germany[97] shows that more than half of people over 80 in nursing homes report having experienced EA. Not only are we horribly mistreating our parents and grandparents, but it is a loss Ultimately, the fate of the economic system hinges on adaptability. Countries that embrace high tech, reform pensions, and balance labor policies could turn decline into a manageable transition. Those countries that don't, or can't, risk stagnation, decline, and a diminished global role. How states will manage these challenges is all to be seen, but, in principle, it is not an impossible challenge in a world where the population decline eases the dramatic lack of resources and

DON'T IDEALISE THE EXTENDED FAMILY.

the damage caused by climate change. Old age is not a sickness; we have to learn how to live with it.

LIVING IN NURSING HOMES ISN'T ALWAYS WORSE THAN A LIVING WITH RELATIVES.

Money is a Gas: Inequality and Population

In 1988, two American economists, Bennett Harrison and Barry Bluestone, were among the first to note the inversion of the tendency toward higher equality in the US that started in the 1960s. Something had placed the country on a path of higher inequality that it is still following nowadays. Most Western countries show the same behavior and some recently industrialized countries, such as China and India, are also seeing an increase in inequality.

But what generated this trend? Harrison and Bluestone proposed several explanations; all related to specific US conditions. Eventually, they concluded that it was the result of a reorganization of the US businesses and corporations designed to maximize profits by attacking labor costs and relocating manufacturing overseas. This is a reasonable interpretation for the US situation, but the fact that the inequality is increasing everywhere, worldwide, indicates that there is something much deeper in this story. I can propose that it is related, at least in part, to the worldwide demographic transition that brought down fertility. But that needs an in-depth explanation.

First of all, what are the causes of inequality? Is it something intrinsic in the way humans are? Or has it to do with the way society is structured? Surely, it is related to money, but how is it that some people have more money than others? In 1976, Pink Floyd created a song titled *Money is a Gas*, which turned out to be a term remarkably apt at describing the distribution of wealth in society according to the "econophysical" approach. It is a vast field, but some studies by Victor

Yakovenko and his coworkers used the "money is a gas" analogy to describe their work on the income distribution in modern times.[98]

To understand their approach in detail, you would need to know something about statistical physics. But the basic idea is simple: they modeled the economy in terms of random exchanges among economic agents. These models are called "money exchange models." The idea is that inequality is a natural result of the statistics of the exchanges, not of special skills or abilities. Money behaves like gas molecules: just like some molecules have more energy than others, some people have more money than others. The models reproduce reasonably well the distribution of money in the real world, where the probability of being rich goes down exponentially with the size of one's stash. In mathematical terms, both money and the energy of gas molecules follow the law called the "Gibbs-Boltzmann" distribution. It is, incidentally, a distribution that maximizes the entropy of the system, making it stable. These results show that a certain degree of inequality is unavoidable in a free market. It may not be a bad thing, provided that it is not excessive.

The rich, however, follow a different pattern, a curve called the "Pareto" or "Fat Tail" distribution, accumulating much more money than the gas model would predict. It is this distribution that gave rise to the commonly heard statement attributed to Vilfredo Pareto, that "20% of the people earn 80% of the income." Why that? Are the rich able to defeat the laws of thermodynamics? No, but they can exploit the fact that there is no such thing as a "free market" in the real world. In the real economic system, people engage in money-exchanging activities not with everyone at random, but mostly with people who are close to them physically and socially. In common parlance, we call this phenomenon "the old boy network." It means that if you have rich friends, you are more likely to be exposed to more opportunities to become rich yourself. In this way, Madame Entropy can be tricked and made to maximize herself only at the local level. That explains the existence and the persistence of the multibillionaires (soon trillionaires) of today.

These models do not take into consideration population growth or decline, but these are very important factors. You know that when you die, you can't take your money with you to Paradise (or Hell, if you have a lot of it). So, death could be a powerful wealth equalizing factor if the inheritance were to be distributed among many heirs. But declining birth rates are changing the rules of the game, shifting the system even more toward the Pareto rules. That is, the rich inherit from the rich, and wealth accumulates rather than being redistributed.

We are seeing a phenomenon similar to the wealth accumulation of the heirs of noble families that was typical of ancient times, and that is described in a previous chapter of this book. The difference is that, in ancient times, rich people had plenty of troubles to make sure that only one of their several heirs would inherit the family's wealth. Today, the problem doesn't exist because many families have only a few children, or just one. No need to cloister anyone; the mechanism functions by itself as a wealth accumulation machine.

The early economists noted the problem. David Ricardo, in particular, noted how landlords, who often inherited large estates, earned "rent" simply by owning scarce land, not through productive effort. Large inheritances created inequality without merit, fostering a class of idle rich who consumed resources without contributing to production, harming economic progress. Later, Socialists often proposed to abolish inheritance altogether, with the wealth of the deceased going all to the state. That, of course, is considered abhorrent nowadays, mainly because the rich see it in that way for obvious reasons.

The Economist magazine rediscovered the problem in an article titled *The New Inheritocracy* on February 27, 2025. It examined the growing importance of inherited wealth in advanced economies and its implications for society. According to the article, the total wealth transfer to inheritance in rich countries was about $6 trillion in 2025. It is equivalent to about 10% of the average GDP of these countries, a value that has doubled in comparison to the mid-20th century. This surge is largely driven by the baby boomer generation passing on wealth, especially in the form of real estate. With the cost of apartments skyrocketing, the gap between heirs and non-heirs is widening in terms of the kind of housing families can afford. If you were lucky enough to inherit a home from your parents, you are in Paradise, and you can dream of your next vacation to Hawai'i. If you didn't, you are in Hell, with a heavy mortgage or rent to pay, and the risk that if you miss a few installments, you'll end up sleeping in Skid Row.

Interestingly, the article in The Economist does not mention low birth rates as a factor in its discussion of inherited wealth. But with fewer and fewer heirs appearing to collect their parents' accumulated wealth, if we wanted to avoid phenomena of extreme accumulation, we would need laws to tax inheritance. Unfortunately, there is no evidence that governments are moving in that direction. We see, actually, the opposite trend. In the US, the federal estate tax remains stable with a 40% rate and an exemption of $13.61 million in 2024. In practice, it affects less than 1% of estates. Some states are imposing additional estate taxes, but most are phasing them out (e.g., Delaware, New Jersey) or are increasing exemption levels. The trend seems to be the same all over the world: many countries are reducing or eliminating estate and inheritance taxes. Even when these taxes still exist, there are usually ways to circumvent them by various legal tricks. It is not surprising: with inequality increasing, the rich are pushing for reducing taxes affecting their wealth, and they surely have more clout than the poor. The process feeds on itself as the rich become richer, and there are fewer of them. Eventually, the poor may not have anything anymore to pass to their heirs.

Given these factors, it seems unavoidable that a shrinking population will lead to increasing wealth inequality. Indeed, it is happening in the countries that experienced the demographic transition. This increase in inequality is often accompanied by the dismantling of the social services that the state used to provide, such as health care, education, social security, and the like. Not only that, but we are also seeing an ideological trend toward assuming that this is the way things should be. The political forces that traditionally defended the poor are in decline or have completely disappeared. Leaders such as Javier Milei in Argentina and Donald Trump in the US represent the trend very well: they want to dismantle the welfare state and replace its services with those provided by private enterprises. Those who can't afford to pay, well, they must learn how to paddle their own yacht, or drown. We can only hope that some trends will slow down if we can find a way to stop the decline and stabilize the population level.

Technology: The Changes Ahead

Typically, we tend to think about technological progress as if scientists were the equivalent of Santa Claus, assumed to bring only good things. But technology may bring coal lumps to everybody, and not just to bad kids. So, in the rush of optimism of the 1950s, people thought we would have flying cars, space colonies, robot butlers, immortality pills, and similar gadgetry. Nobody could imagine things like the Internet, social media, genetic engineering, photovoltaic energy, and the like. Nor could we have imagined the disasters that these wonderful technologies would bring in terms of pollution, climate change, erosion, ecosystem disruption, loss of biodiversity, and more.

The only rule that we can extract from about a century of technological forecasting is that the technologies that are most touted are those that tend to fail. A good case in point is that of the "hydrogen-based economy," which governments doggedly continue to push and finance, but which never worked and probably never will. But new technologies are having big effects on many fields of the world's economy. What will be their effect on the human population trends? Let's see a list of what we may expect, with the elements in no particular order.

Artificial Intelligence. AIs have burst onto the scene during the past few years, making human work obsolete in many fields. In terms of human population trends, AIs cannot directly affect the production of food or human fertility, but they may affect the economic system in various ways, mainly because they will change the value of human beings as workers and soldiers. If humans become less valuable, there may be a lower interest in increasing the population, and that may lead AIs to become powerful population-limiting factors. A further possibility is that people will become so addicted to AIs that they'll lose interest in everything else, including sex and reproduction. "Sexbots" may become powerful birth control factors, enhancing the already existing ones. It is also possible that AIs could turn out to be not so friendly as they seem to be nowadays. And that would mean bad news for us, puny humans.

Renewable Energy. The Industrial Revolution started during the late 18th century as the offspring of fossil fuels. Coal first, then oil, and later natural gas provided the energy to move the giant machine we call the "economic system." By now, fossil fuels are obsolete; solar photovoltaic and wind energies are revolutionizing energy supply at costs vastly inferior to what fossil fuels can produce. That promises the capability of staving off the collapse caused by a lack of energy, which is still believed to be coming. Renewable energy can power the many processes of agriculture and, in principle, help generate abundant food for human beings. The problem is that building a new energy system based on renewable energy will take time and money, and we seriously risk an energy crisis before the transition is completed. In addition, we are currently seeing a propaganda attack against renewables, likely supported by the fossil industry. This attack is turning renewable energy into anathema for many people, especially on the right side of the political spectrum. Not for nothing, President Trump said that renewable energy is a scam and a "con job" in his 2025 speech at the United Nations. It may be that before we understand how much we really need renewable energy, we need a Seneca Cliff in energy production that will sweep away fossil fuels from the energy system. That, unfortunately, won't be painless, and it may generate a parallel Seneca Cliff of the world population.

Lean Technologies. The Industrial Revolution was ushered in by an abundance of mineral resources. Not just fossil fuels but all the elements of the periodic table have been mined and lavishly used as if there were no tomorrow. Now, the industry is facing a dearth of essential minerals, e.g., copper for conductors, rare earths for magnets, noble metals as catalysts, and more. This trend is forcing a complete rearrangement of the industrial system according to the "three golden rules of mining," which I sketched in my 2014 book *Extracted* (2014). Inspired by the working of the biological ecosystem, these rules are 1) Use only what's abundant, 2) Use sparingly, and 3) Recycle ferociously. By using these rules, Earth's ecosystem has been thriving for a few billion years without ever running out of anything. Unfortunately, that's not the way the human economy works: it is wasteful, inefficient, and scarcely interested in recycling. So, the risk of running short of critical minerals is not negligible. Nevertheless, current critical technologies can be re-engineered to work according to these rules. For instance, PV panels use elements abundant in Earth's crust: silicon and aluminum. At the same time, the transmission of electrical energy can be done by replacing the rare element copper with the abundant aluminum. Mineral depletion is a problem that population stabilization and decline can substantially ease.

Genetic Engineering. Theoretically, hacking the DNA of living beings opens immense possibilities. In terms of the effects on the human population, one of their potential uses in medicine is to lengthen the human lifespan. So far, however, the global trend is tapering off worldwide at about 73 years of life expectancy at birth. Even the nations that have the longest lifespan, Italy and Japan, have seen their life expectancy no longer increase. The "pill of immortality" that was envisaged in the 1950s doesn't seem to be coming, just as we don't have miracle cures for sicknesses such as cancer, dementia, diabetes, and more. The miracle cure for obesity, Ozempic, is now being adopted everywhere in developed countries. However, it has the problem that once people start taking it, they can't stop; otherwise, they will become obese again. And we are not sure of the long-term side effects of this drug. Yet, this field is in motion, and it is still possible that new technological developments could substantially increase the human lifespan. In this case, there would be consequences on the global population curve, with the population peak being delayed and the social structure being even more dominated by elderly people.

A society dominated by the elderly ("senocracy" or "gerontocracy"), is a condition never experienced in the past. It was only described in fictional form by Jonathan Swift in his 1726 book, Gulliver's Travels. In the story, the *Struldbrugs* are people born with a red spot on their forehead who are immortal but do not have eternal youth. By the time they reach old age, they are decrepit, miserable, senile, in constant pain, unable to enjoy anything, and legally dead (they lose all civil rights at 80). They become an unbearable burden to themselves and society, envying the ordinary people who are lucky enough to die. Is this our dark destiny? Maybe it is, considering Fran

Lebowitz's quip that "my objective is to become decrepit, because the alternative is worse."

Robot Soldiers. Robots, under the name of "drones," are revolutionizing many fields, including warfare. In a chapter I wrote for the book *2052 –A Global Forecast for the Next Forty Years* (2012), I discussed the consequences of drones replacing human beings on the battlefield, arguing that it wouldn't be a bad thing. We are seeing it happening right now in Ukraine. Drones haven't yet made infantry completely obsolete, but we are moving in that direction. So, are we going to see a human-free battlefield in which drones will fight each other in a sort of huge ball-wrecking game? Something similar to the British TV series, Robot Wars, aired from 1998 to 2004 and from 2016 to 2018. It featured clumsy robots fighting each other with hammers, circular saws, and other mechanical devices. Why people found that fun is difficult to explain, but at least it was a harmless pastime.

In any case, the idea that robots will fight for us is at least a bad waste of resources and energy, but all wars are like that. The real problem is that robots may also be programmed to fight human beings, and that would provide a cheap and effective weapon of mass destruction. We can only hope that they will remain under the control of sane human beings.

The New Green Revolution: Precision Fermentation. The conventional Green Revolution was a set of technologies mainly based on mechanization, fertilizers, and pesticides, in turn, all based on the availability of cheap fossil fuels. This approach has now exhausted its capability for further improvements. But if we move to precision fermentation powered by renewable energy, then, in principle, it would be possible to increase food production and sustain a much larger population than the current one. The result would be humankind further expanding into the ecosystem and leaving even less space for non-human creatures. It is not impossible to conceive such a dystopic scenario, reminiscent of Asimov's short story, *2430 A.D.* (1970). Philip Dick also explored this theme in his *Do Androids Dream of Electric Sheep?* (1968), which was the basis of the *Blade Runner* movie (1982). The common theme of these stories is that humans would increase their numbers so much that they would cause the extinction of most animal species. Some people seem to think that this is a good thing, such as some religious fundamentalist groups and a loose group of Silicon Valley thinkers, transhumanists, and futurists (e.g., Ray Kurzweil, Peter Diamandis, and others). It is, after all, what's happening right now with a new mass extinction in full swing in Earth's ecosystem. On the other hand, even assuming that technological developments could feed several tens of billions of people, there are stumbling blocks that could make it impossible. One is that humans still need a functioning ecosystem to which they are adapted as a result of millions of years of evolution. A one-species, human-only ecosystem may be simply impossible to maintain in stable conditions. It could crash because of pollution, lack of nutrients, or perturbations we can't yet imagine.

These considerations describe some of the main trends of technological progress, those which are likely to have a significant effect on population trends. But the situation is complex, and one characteristic of technological progress is that it always takes us by surprise. It could lead the human population to tens, or perhaps hundreds, of billions. Or, conversely, it could rapidly lead to the extinction of the human species. Unfortunately, the future will not be what we think it should be.

Global Warming: The Ultimate Threat

The current debate on population growth often deals with the question of whether a population decline will reduce emissions and save humankind (and the ecosystem as well) from the grim reality of global warming. It is grim because Earth's atmosphere has a "memory" of the past; it remembers the accumulated emissions in the past. So, we don't just need to stop emitting greenhouse gases, but we should take steps to reabsorb the already emitted amounts.

Unfortunately, the current political ideas are not favorable to an effort to reduce emissions, to say nothing about doing something to restore the atmospheric composition to the pre-industrial characteristics. On the contrary, many governments are pushing for an increasing use of fossil fuels, while major ones openly define the global warming problem as a "scam," as the US President, Donald Trump, did.

Given the situation, could the expected decline in humankind's population reduce the warming threat? Fewer people mean lower emissions, and that's a trend that goes in the right direction. In practice, though, the question is extremely complex and involves a large number of parameters having to do with accumulated climate factors and the complex way in which the Earth's ecosystem is reacting to the perturbation. There have been only a few studies on how population decline could affect Earth's climate, probably because the idea that the global population is close to its peak has not yet penetrated the general perception of the scientific community engaged in climate science studies. Typically, climate models only tell us how much temperatures will increase in dependence on greenhouse emissions and, in most cases, emissions are simply expected to increase with population, unless they are curbed by internationally agreed legal actions.

Only a few studies have explicitly attempted to quantify the consequences of population trends on climate. These models are called "Integrated Assessment Models" (IAMs) and try to describe not just the climate system, but the coupled economic and climate system. A recent review of these models can be found in the paper by Krawczyk and Braun.[99] The problems with these models are many. Krawczyk and Braun describe the "Superhero Syndrome" that affects IAMs, that is they can

> "… be compared to Hollywood superheroes, fixating on defeating one villain, like emissions, while overlooking the bigger picture. Despite their heroic efforts, IAMs and superheroes overlook the broader systemic issues, and thus they mark the prevailing crisis-ridden societal conditions as rational and legitimate. In other words, IAMs examine futures to find solutions to mitigate the climate crisis, but the way they do it stabilizes the status quo and thus the path into the climate crisis. "

My personal experience with these models is based on the MEDEAS one, which I helped develop,[100]It was a complex and powerful model but, in my opinion, the large number of parameters involved in the computation made it subject to biases and uncertainties. This problem exists for all large and complex models. Conversely, some IAMs are oversimplified and unable to provide reliable results. This is the case of the first of these models, the one called DICE (Dynamic Integrated Climate Economy) model, which is today the most common and most influential among IAMs. Unfortunately, this model suffers from serious biases and problems.

The DICE model has its origins in the work of William Nordhaus. In the 1970s and 1980s, he focused on modeling the effects of mineral depletion, criticizing the results of the world models generated by system dynamics, such as The Limits to Growth. I describe this approach in detail in my book *The Limits to Growth*

Revisited (2011). For this purpose, Nordhaus used a simple model based on the concept of "production function" and, taking into account the effect of technological progress, argued that the global economy would keep growing exponentially for the foreseeable future. Later, Nordhaus introduced global warming as a parameter of the model. The result was the DICE model, proposed for the first time in 1992,[101] and later upgraded in various versions. On the basis of his model, Nordhaus and others typically arrived at the conclusion that climate change would only marginally affect the economy, and that mitigation efforts should not be undertaken at levels high enough to reduce economic growth. Nordhaus was awarded the Nobel Prize in economics in 2018 in large part because of this work.

Here, it would be out of place to enter into a detailed discussion of the DICE model. Briefly, anyway, it is based on the assumptions typical of the neoclassical economic approach. The output of an economy is proportional to the product of two main parameters, capital and labor, affected by an entity called "total factor productivity," which is supposed to represent the exponential increase of technological progress. Population is an exogenous parameter; that is, it is not determined by the inner parameters of the model, but by external projections generated, for instance, using the UN demographic models. Emissions are proportional to the economic output, and are supposed to decline via carbon-saving technologies. Temperature is also an exogenous parameter, supposed to vary according to the IPCC estimates.

The limits of this approach are many. For instance, increasing temperatures are not supposed to cause extreme events such as hurricanes or generate disastrous "tipping points" such as the melting of the polar ice caps or the collapse of the Atlantic Meridional Overturning Current (AMOC). Additionally, rising temperatures do not directly affect population growth at all. The only effect of Global Warming in the model is on the GDP, assumed to face a decline proportional to the square of the temperature increase (or "temperature anomaly" in the climatologists' jargon).

Apart from being an arbitrary assumption, the effect of temperature appears to be badly underestimated in the DICE model. Just consider that a temperature increase of 2-3 °C only generates a loss of a few percentage points of the world's GDP, which appears overoptimistic, to say the least. Larger increases, of the order of 6 degrees or more, have no disastrous effects, apart from a modest decline in the global GDP. In addition, the fact that the damage function increases with the square of the temperature anomaly naturally leads to postponing efforts to mitigate the effect to the far future, when the damage will be really serious. The same is true for the steep "discount rate" embedded in the model, which evaluates long-term damage at a lower cost than short-term ones. That encourages postponing investments until the problem can no longer be ignored.

How the consequences of the DICE model are understood by economists is spelled out clearly in the book by Dean Spears and Michael Geruso, *After the Spike* (2025). On the basis of a recent version of the model, they extrapolate that the human population will grow to about 10-12 billion people at the end of the century, then it will stabilize itself, or start a slow decline. The GDP will continue increasing, leading to a richer world. Temperatures will not increase above ca. 4 °C, and that will cause no disastrous damage.

The gist of the idea of Spears and Geruso is that a large human population will have a higher economic output and will be able to develop new technologies and field a larger climate change mitigation effect in the future. As a consequence, no action should be attempted now to face a problem that is expected to be easier to solve in the future. The system will optimize itself to a high level of prosperity for everyone. In short, it is a view based on something similar to the "invisible hand" concept introduced long ago by Adam Smith, now supposed to allocate resources in the optimal way to solve the climate problems while maintaining economic growth.

Unfortunately, the DICE model is not only overoptimistic, but it also lacks the features that would be necessary for a realistic representation of the effects of rising temperatures on the various elements of the ecosystem and of the human economy. That is, it lacks the effect of *feedbacks.* Earth's climate is not a static entity that we humans just happened to find when we evolved. It is part of a complex system dynamically controlled by the biosphere. It can react to perturbations by abrupt changes.

Think, for instance, of the possibility that in some regions, temperatures will become too hot for human beings to live in. North Australia, central India, and the Persian Gulf areas are all places where temperatures are expected to exceed the "Wet Bulb Temperature" at which people die of thermal shock if they stay in the open for more than a short time. But you don't need to arrive at that point to experience famines caused by food production decline, or lack of water access, or simply living a miserable life in summer if you can't afford air conditioning. And don't forget extreme events: you and your home may be swept away by hurricanes, and the rising water level will make it easier for tsunami waves to reach you. Wildfires are another problem enhanced by increasing temperatures. Finally, you need to take into account that carbon dioxide is not just a climate forcing agent. It is a chemically active substance active in the metabolism of the biosphere, including the metabolism of human beings. In some cases, it may be seen as "food for plants," but for those of us who are not plants, it can have all sorts of harmful effects, including a decline in cognitive capabilities.[102]

Unfortunately, while a lack of models is a bad thing, wrong models are even worse. The DICE model and its subsequent version may be leading decision makers into a false sense of security by generating scenarios that promote economic growth in place of mitigation. If Earth's ecosystem goes through one of these points, say, the collapse of the Gulf Stream, it will no longer be a question of maintaining economic growth; it will be a question of survival for humankind as a whole.

So, abandoning the optimistic DICE model, what can we say about the effect of a slow population decline on climate? World models based on system dynamics do not normally have global temperatures as a parameter. Nevertheless, from the start of the field, in the early 1970s, these models included a parameter called "pollution," or "persistent pollution."[9,10]. At the time, the threat of global warming was not completely understood, yet, but later it became commonplace to consider the pollution stock as a "proxy" for the concentration of CO_2 in the atmosphere.[111,112]

System dynamics models can therefore tell us what to expect will happen to the CO_2 atmospheric concentration and, hence, the future global temperature trends. We cannot expect these models to have the same degree of refinement as the current climate models, but they can give us at least some hint about what to expect as a result of the interplay of the various parameters Let's skip the results of the early studies and focus on a relatively recent version of *The Limits to Growth: the 30-Year Update*, published in 2004.[33]

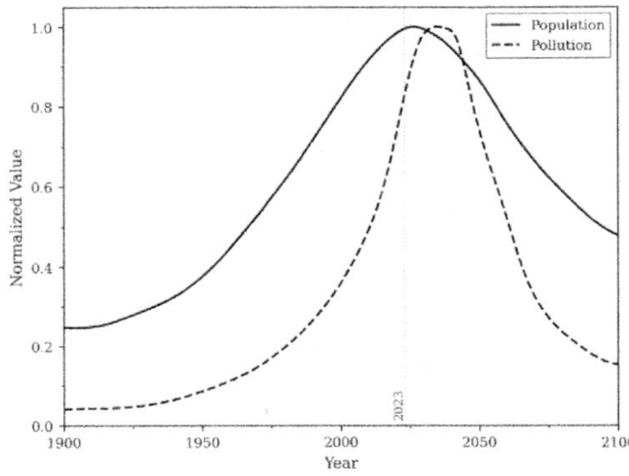

Figure 39 -- The data for population and pollution parameters from the "base case" scenario of The Limits to Growth, the 30 years update (2004)

The figure shows the data for two of the parameters, population and pollution, for the "base case" scenario of the 2004 study, that is, the one based on the data available at that time and which assumes no action to steer the economy in the direction of a lower use of resources or reducing pollution. In this calculation, the global population peaks in ca. 2027-2028, pollution peaks about 10 years later, and then rapidly goes down. This is a scenario that describes a future we may see as not so bad. In these assumptions, the gradual decline of the world's population may save us from the worst damage generated by climate change. Note, though, that it won't save us from a bad crash of the economic system generated by the depletion of natural resources.

Unfortunately, this nice scenario is the result of a specific choice of parameters, and we can't expect the universe to make the choices we would like it to make. Let me show you another scenario from the same study. This is called "BAU2," and it assumes that the available natural resources are twice as large as in the BAU1 scenario. It is equivalent to assuming that better technologies will allow the industrial system to use resources more efficiently, or to find new ones

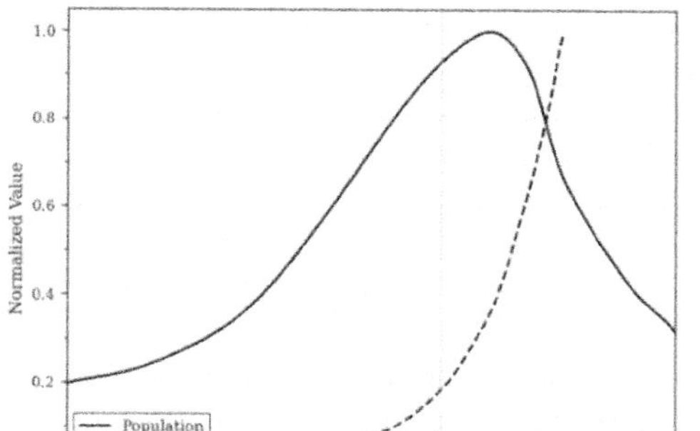

Figure 40 - The data for population and pollution parameters from the BAU2 scenario of The Limits to Growth, the 30 years update (2004). This scenario assumes a higher availability of natural resources.

In the BAU2 scenario, the population peaks around 2048 and then crashes down in the typical shape of the "Seneca Cliff," while pollution shoots through the roof. If we see the pollution parameter as proportional to the concentration of CO_2 in the atmosphere, as it is commonly done in these studies, it would bring Earth back to a situation comparable to that of the End-Permian catastrophe that led to the near extinction of the biosphere.

These calculations were made more than 20 years ago; we have a more recent version with the study by Nebel et al. [28] who recalibrated the Limits to Growth base case scenario, adapting it to the data available in 2023.In the 2024 recalibration, the global population peaks in the 2030s. Pollution keeps growing for at least 30 years afterward and starts declining only after ca. 2070. Considering that the pollution level at the peak is about three times larger than it is now, we might estimate that in this scenario, the CO_2 concentration would rise to well above 1000 ppm, generating a temperature rise of 5-6 °C. A condition that humankind would probably not survive.

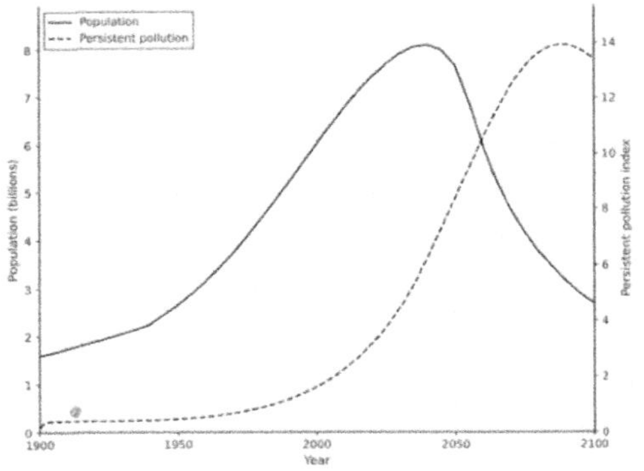

Figure 41 – The data for population and pollution parameters according to a recalibration of the WORLD3 model performed in 2024 28

This romp on modeling should be useful to understand the basic features of the global warming threat. For one thing, the uncertainty is large when projecting a vast and complex system such as Earth's climate into a long-term future. You saw how different models and a different choice of parameters could lead to scenarios that go from a world similar to the scorched Earth of the early Mesozoic to a world where only minor adjustments need to be made in order to keep the economy growing. There are many more models, and you can play with your own. But only the Universe knows which is the right model and what the right parameters are.

Unfortunately, uncertainty is often used as a tool to favor financial or political interests. It is a propaganda technique that has been known for a long time, described in detail in the book *Merchants of Doubt* (2010) by Naomi Oreskes and Erik Conway. It was extensively used by the tobacco industry in the 1950s, during the debate on the health effects of smoking. If you have been involved in debates on climate and similar subjects, you surely encountered the attitude that goes as "since the damage that global warming creates is uncertain, then we should do nothing until we know better." In many fields of life, this attitude would be labeled as criminal. Imagine someone telling you, "Since it is uncertain whether you'll have a car accident or not, you shouldn't have your child wear seat belts." We react correctly to this attitude when it deals with things we know well, but climate change is sufficiently remote from our everyday experience that the "uncertainty" technique is effective.

But uncertainty is not a bug; it is a feature of models. Nobody can predict the future, and prediction is not the job of models. Their purpose is not to tell us what the future *will* be; it is to tell us what the future *could* be. Just as you can't forget that car accidents happen, you can't forget that abrupt bursts of increasing temperatures happened in the geological history of our planet. And if they happened in the past, they could happen in the future. What we have done so far has been sufficient to unbalance the system enough that it might go through one of those transformations and bring us to another world. It is the scenario that Bill McKibben named *Eaarth* in his 2010 book with that title. The models tell us a few fundamental points; the main one is that we can't expect population decline to solve the climate problem automatically. But it can also be of considerable help to reduce the damage. The destiny of humankind lies between these boundaries.

Reversing the Fertility Decline?

Population decline is not a bad thing in itself. We may discuss its disadvantages, but a positive feature of it is that it can reduce the impact of humankind on a heavily strained global ecosystem. It means having more resources available per person and less pollution overall. But there also exists the concept of "too much of a good thing." Without input from new generations, states and societies could decline, or even collapse, unable to maintain the social and economic structures they had been able to build with the contribution of a larger population. So, could the future population trend maintain the good things of our society, including social security, health care, public education, and the like? Ideally, the fertility could stabilize at a level that ensures generational exchange, while slowly having the population decline to levels that the ecosystem can support. The process should be slow enough that it could maintain a working economy.

Could that happen? The answer depends on the factors that affect fertility. We saw in this book that there are at least four factors involved in controlling population size: famines and wars, socio-economic factors, pollution, and stress. Fortunately, right now, famines and wars are not playing an important role, while stress doesn't appear to be significant, either. But we still have to take into account socioeconomic factors and pollution. It is unlikely that the fertility will adjust itself to the ideal levels we would like it to be, so we need governments to intervene on these factors: creating socioeconomic conditions that encourage families to have children, and abating pollution in such a way as to make it possible for them to do that. In this way, the population could be stabilized without the need to force anyone to have or not to have children.

Let's start by supposing that the fertility decline is mainly the result of socio-economic factors. That is, fertility declines because people tend to manage their family size in order to maximize their well-being and that of their children. In this case, fertility could be regulated by interventions in the socio-economic sphere. Financial support, facilities to take care of children, bonuses for working mothers, and other forms of help for families with children are all good things that could lead to a return to higher birth rates. This idea is already being tried in several countries in various measures. The problem is that it appears to be ineffective: birth rates are not increasing.

A compendium of the current data on this approach can be seen in a recent article by John Burn-Murdoch in The Financial Times of March 29, 2024, appropriately titled "*Why family-friendly policies don't boost birth rates*." Between 1980 and 2019, the world's most developed countries roughly tripled their spending on child benefits, but they all saw birth rates decline, sometimes at increasing speeds. In addition to the cases reported in this article, we may cite that of Russia, which is implementing several measures to encourage women to have more children, so far with meager results. There are no examples, so far, of a country where government intervention to favor natality led to a consistent reversal of the declining trend.

Maybe governments didn't pay enough? Indeed, the government subsidies tried so far are paltry sums in comparison to the cost of raising a child to adulthood. Let's consider the case of the US, where a middle-income, married-couple family with two children is reported to spend about 200-300 thousand dollars (USDA data) to raise one child from birth to age 17. This is necessary in terms of housing, food, childcare, education (pre-college), healthcare, transportation, and more. It equates to about $15-20 thousand per year per child, up to age 17. In practice, in most cases, the family will continue to support their children for several more years. We also know that in the US, there are about 80 million underage people. Assuming a yearly cost of $20,000 per year, the total is roughly $1.5 trillion per year. It is a huge sum. The U.S. federal budget for fiscal year 2024 was approximately $6.8 trillion. A $1.5 trillion extra subsidy for underage people would represent about a quarter of that total. Subsidizing families for 100% of the cost of raising a child would only be possible by raising taxes or cutting other government expenses by the same amount.

In the current political climate, raising taxes at the necessary levels is simply unthinkable. About cutting other government expenses, not even the budget of the mighty DOD, the Department of Defense (now DOW, Department of War), would be enough; it is today under one trillion dollars per year. And, in any case, the idea of using a large fraction of the military budget to finance child raising is even more unthinkable and unspeakable at a time when the development of drones has made human cannon fodder useless. These results are for the US, but wouldn't be very different in other industrialized countries.

Even if the government were able to increase support for childbirth and child raising, would it solve the depopulation problem? Never forget the Easter Island fallacy: "Our problems will be solved if we just build bigger statues." In our time, it is expressed as "if we throw enough money at the problem, eventually it will go away." The thorny issues with child-raising subsidies are many, not just that the amounts that can be reasonably paid are too small. How to distribute subsidies? For what specific purposes? Subsidies might overcompensate some families (e.g., the lower-class ones, which are spending less per child) and undercompensate upper middle-class ones. To say nothing about racial, religious, and ethnic issues that would put different sections of the population in conflict with each other, and to grab subsidies that would allow their groups to expand in numbers at the expense of other groups.

Seen from a different viewpoint, large subsidies would amount to paying women to have children, which would give a new meaning to the concept of "proletariat" of Marxist times. We might think of some extreme scenarios in which government subsidies would bring about a profound restructuring of the social order, with some women taking childbearing as a profession. At the limit, society could organize itself in forms resembling the story told by Margaret Atwood in her novel *The Handmaid's Tale* (1985). It describes a dystopian society based on a distorted reading of the Bible, where fertile women are enslaved in order to produce children for the state. Society would then turn into a "eusocial" arrangement in which reproduction is a task taken by a small number of females, as is the case for some insects such as ants and bees. Some mammals adopt this arrangement, for instance, the naked mole rat (*Heterocephalus Glaber*), a creature native to the Horn of Africa and parts of Kenya. Those small creatures are said to have developed this social structure in order to survive the extremely harsh conditions of their environment. Maybe we will have to learn something from those rats? Right now, it looks unthinkable, but the recent years have seen so many previously unthinkable things becoming obvious truths that we can't completely discount anything.

A less complicated strategy to increase fertility could be the stick instead of the carrot. That is, penalizing people and families without children. This is a known approach, traditionally applied to men. In the past, several countries enforced "bachelor taxes" designed to encourage marriages. In Italy, the Fascist government introduced a tax on celibacy in 1927. During Nazi times, Germany introduced taxes to promote the "Three Ks," *Kirche, Küche, Kinder* (Church, Kitchen, and Children), which were supposed to be the main occupations of women. Probably not exactly what we want in the West today.

In any case, there is little evidence that these methods had any effect, but we cannot exclude the possibility that they will be tried again. They have the advantage that they cost nothing; on the contrary, they bring money to governments. But forcing people to marry is one thing; forcing them to have children is another. Taxation on childless people would be unfair for those who can't have children for health reasons, in addition to bringing many more social and political problems.

Another method to raise fertility could be to use the best weapon that governments have to convince people to do things they don't want to do: propaganda. Having children could be praised and presented as something that provides prestige for the women who engage in the task. In the 1950s, it was common to present happy families with four to five children in advertising. In Eastern countries, natalist propaganda had a more socialist slant, but it had similar themes. The perfect Soviet mother was termed a "heroine" and was sometimes shown with 8-10 children. As it is typical of propaganda, these ideas faded out with the stark reality of the high cost of having children becoming clear to families. This method may well be tried again, but it is unlikely that it will have a large effect.

A different approach to propaganda is based on demonization. You probably know that, as Machiavelli never said, the art of politics consists in finding a culprit. So, this culprit could be found in childless families, working women, women who have a culture, and similar targets. Working women would lose their jobs, girls would be discouraged from attending school, illiterate women would be praised, while cultured women would be shunned and considered evil people. That would lead to a situation not unlike the way witches were demonized and despised in ancient times. Could society arrive at a new round of active witch-hunting to get rid of those pesky women who insist on thinking with their own heads? Again, it is an extreme scenario, but in recent times, we have seen the unthinkable becoming perfectly thinkable, and the current debate on social media seems to be moving exactly in that direction.

What if, instead, the fertility decline is mainly the result of pollution? In this case, the problem is much more difficult in political terms. Whereas laws and financial subsidies are the normal terrain of action of governments, doing something against pollution is more difficult and more expensive. One problem is that it goes against the sacred principle of economic growth. Another is that polluters have powerful weapons to influence government choices that damage their profits: propaganda, lobbying, and good old corruption of government officials. The strong resistance of the polluters' lobby may be one of the main reasons why the idea that pollution is the main cause of fertility decline is mostly ignored nowadays. Nevertheless, there are several examples of successful policy actions designed to curb pollution. It has been possible to eliminate at least some major deadly pollutants: lead, cadmium, beryllium, CFCs, DDT, and others. A much-touted example is that of the elimination of the chlorofluorocarbon (CFC) compounds that were destroying the ozone layer in the upper atmosphere.

The problem is that in the current situation, the government engagement should be stepped up to cover much more dangerous and common pollutants known to cause infertility. That is, plastics, which are the main source of endocrine-disrupting molecules such as BPA, PFAS, pesticides, and more. Then, we should also curb the production of heavy metals and all those industrial activities that indirectly affect human health, primarily the emission of greenhouse gases.

Unfortunately, fighting pollution is expensive and unpopular, especially in the current political climate. The public perception of the pollution problem appears to have gone full circle: from the indifference of the first half of the 20th century to the worries that started being common in the 1960s. Nowadays, we see a return to neglecting the problem, while those who are raising the alarm are demonized and ignored. A good example of the trend is how the recent turn to right-wing policies is working at undoing regulations that were protecting people. Think of the recent *Plan 2025*, prepared by the Heritage Foundation, a prominent conservative think tank, in partnership with over 100 conservative organizations. It is said to be a manifesto of the current right-wing views and, among many other things, it contains the specific proposal to "revisit the designation of PFAS chemicals as 'hazardous substances.' In practice, that means to deregulate the PFAS concentration in drinking water and food, throwing away all the evidence that it causes not only infertility but also serious health damage to people. Plan 2025 says nothing about other endocrine-disrupting substances, probably because the authors didn't know they existed. But there are other examples of failures in the attempts to control pollution. One is the recent attempt to remove carbon dioxide, CO_2, from the list of polluting substances.

Do these people drink the same water and breathe the same air as we do? Do they think they are invulnerable? Are they human beings? Or maybe Reptilian Aliens? More simply, these people may genuinely believe that they have been bothered enough with all those sacrifices that governments have been asking us to make to combat climate change and pollution. That is, we are burdened with expensive requests, unable to solve the problems they claim to solve – assuming that these problems exist. According to this view, the idea that PFAS is bad for people's health is just another scam perpetrated on us by those dishonest scientists who want to get money for their research grants.

With the current political turns that emphasize profit at the expense of protecting people from pollution, the possibility of significantly abating the concentration of endocrine-disrupting substances and other pollutants appears slim. Eventually, though, a declining population will also cause the decline of pollution, since industrial production will decline. Unfortunately, scenarios such as those calculated by *The Limits to Growth* series of studies tell us that worldwide pollution will start to fall decades after the collapse of the industrial system. Hence, we are destined to live with the pollution we created in the past, and we are still creating.

Can the fertility decline problem be solved by technological solutions? Much can be done in this sense, but new technologies typically work on symptoms rather than on the sources of the problems. Regarding fertility, there are no approved medical treatments to eliminate endocrine-disrupting substances from people's bodies and restore fertility and sex drive in people who have been heavily exposed to them. At present, the only technology that can reverse the fertility decline is in vitro fertilization.

"In vitro" is a term that comes from the Latin word *vitrum*, which means glass or crystal. It is a method that begins with hormonal stimulation of the ovaries of the woman to produce multiple mature eggs, which are then surgically retrieved. The eggs are fertilized in a laboratory dish by sperm cells obtained from a male donor. When embryos form, one is selected and transferred into the woman's uterus, then a normal pregnancy ensues. The method bypasses all the problems that the poor little sperm cells have with having to swim all the way along the fallopian tubes. It also overcomes the decline in sperm cell numbers: only one of the tiny creatures is needed to fertilize an egg. In vitro fertilization is already relatively common in industrialized countries. According to the American Society for Reproductive Medicine (ASRM), about 98,000 children were born in this way in 2023, and the number has been increasing from the values of the previous years. It represents 2.7% of all U.S. births, out of about three million and a half in total. Not a large fraction, but not negligible, either.

Could the method be expanded to have a more significant impact on fertility rates? In principle, yes, although at a high cost. At present, a single fertilization cycle goes for about $20,000 in the US. For women under 35, in vitro live birth rates per cycle are around 40-50%. Over 40, it drops to 10-20%. So, multiple cycles are often needed, and the cost rises in proportion. However, prices may go down in the future if in vitro fertilization becomes common. Already now, they are much lower in some countries, and, among other possibilities, a woman may try Ukraine, where one cycle of in-vitro fertilization is reported to cost less than $3000. So, it is perfectly possible that in vitro fertilization will become a common way to give birth to children, perhaps even the most common one if the current trends continue. After all, one hundred years ago, the idea that most children would be born in hospitals would have been nearly unthinkable. So, in the future, it may become perfectly normal that nearly all children will be created in a glass vial.

Even if we arrive at an extensive use of in vitro fertilization, it doesn't mean that all problems will be solved. Some factors that damage natural fertilization remain; for instance, the quality of the DNA carried by the sperm and egg cells. The main questions, though, are social and economic. In the technique known as "gestational surrogacy," the woman carrying the pregnancy need not provide the egg, allowing her to serve purely as a "womb for hire." This role is indeed a paid service. Sperm, meanwhile, may come from the mother's partner or an external donor, who might also receive payment as a sperm provider. These possibilities carry substantial additional costs for assisted reproduction, making it accessible mainly to the rich. In this case, in vitro fertilization would reinforce the current socioeconomic trend of having fewer children by the average population.

A deeper danger associated with these technologies would be that they can pave the way for eugenic practices once deemed impossible. For instance, with in vitro fertilization, it is possible to determine the sex of the embryo before it is implanted. Hence, parents can select the sex of their child without the need to resort to abortion. That could imbalance the sex ratio in those cultures where male children are preferred. This method is illegal in many countries (although not in the US), but we all know that laws are not always respected.

Another new possibility offered by in vitro fertilization is screening the fertilized embryos for potential genetic defects and discarding those that would produce defective babies. The technology has arrived to the point that the embryos can be screened by sophisticated DNA analysis techniques in order to choose which embryo is the most likely to produce a healthy baby, unaffected in late life by common forms of illness, such as high blood pressure or diabetes.

The possibilities offered by in vitro fertilization are multiple, and not all positive. For instance, governments could impose which eggs and sperm can unite in a Petri dish under laws favoring genetic optimization. What was once a private choice of partners could become a state-orchestrated selection; a dystopia we rightly view as abhorrent today. Yet history tells us that when something becomes possible, then it may well become reality. We all know how, in the recent past, there have been governments engaged in eugenic procedures to improve human beings of selected races. Could these ideas return? Hopefully not, but we can't be sure about that.

So, what is the likelihood that population decline may be slowed down without resorting to extreme technological or social measures? The best we can say is that governments should be careful to avoid worsening a situation that may be worrisome, but not yet catastrophic. There is no reason to panic: population decline is a natural phenomenon that will not lead to the extinction of the human species. First, even though we don't know whether pollution is the most important factor causing fertility decline, cleaning up the environment as much as possible is a good idea for many reasons. Cleaning up the air we breathe, the water we drink, and the spaces in which we live will surely improve people's health and will also reduce the societal costs for health measures.

We also need to help families to have and raise children, without expecting them to return to the fertility levels of the times before the Industrial Revolution. No one could, or should, be forced to have children, or enticed to have them by propaganda or other tricks. If we can achieve a clean, harmonious, less stressful, and more balanced society, it is perfectly possible that fertility will return to the replacement level that we need to keep our society alive and well.

Beyond Earth's Limits or Extinction?

Discussing the global human population trends often starts with the question, "How many people can Earth support?" The implication often is that we are already too many, and sometimes that it could support many more. But the question doesn't make sense unless a condition is specified: for how long? There is no doubt that the current global food production system can support eight billion people because it is doing exactly that. But how about the situation in a few decades from now, with fertilizers and pesticides becoming more expensive because of fossil fuel depletion? How about the effects of global warming, with the associated droughts and extreme events? How about pollution reducing fertility to levels that cannot sustain the human population? It is clear that "population" is a dynamic element of our future: it can, and most likely will, change, increasing or decreasing depending on human actions and on how the ecosystem will react to the increasing human pressure. The only thing we can be sure of is that the future is never what we expect it to be. Nevertheless, the point of reasoning about the future is to be prepared for it. So, what can we reasonably expect?

There are plenty of extreme population scenarios floating on the Web. On the pessimistic side (at least from the human viewpoint), we have extinction. It could hardly be caused by a single factor, but rather by a combination of events, including famines, plagues, wars, and assorted disasters. Climate change alone could make Earth uninhabitable by human beings, and that would conclude the human adventure on this planet. Extreme, indeed, but not impossible. After all, it is the normal destiny of species to go extinct. Why should humans be an exception?

Less extreme views do not imply extinction, but a return to conditions typical of ancient human societies. The "Olduvai Scenario," proposed by Richard Duncan,[103] posits the return to hunting and gathering. Other scenarios foresee the return to a purely agricultural economy, similar to the conditions of Europe during the Middle Ages. In both cases, a regression to ancient times would be caused by the increasing cost of mineral extraction, which would make it impossible to sustain an industrial civilization. Both scenarios imply a huge reduction in the human population, considering that it had always been well under one billion individuals before the Industrial Revolution started in the late 18th century.

On the opposite side, we have optimistic views that see technology propping up human expansion, overcoming all the doomsday views that have been fashionable during the past few decades. Maybe we will find an energy source truly "too cheap to meter," as was dreamed of in the 1950s. Then, we could couple it with planetary engineering in such a way as to control the atmosphere's composition and temperature. Using artificial food, we could decouple humankind from Earth's limited resources. At this point, we could colonize the Moon and Mars; terraform Venus, and colonize it too. If you really trust technology, we could colonize space and build huge orbiting habitats around Earth or other planets.

We could even build a "Dyson Sphere," a huge shell around the sun that would collect all the light that the sun emits and use it to create a biosphere on the inner surface. That's beyond what current or even imaginable human technologies could do. Still, if it were possible, it would provide the definitive answer to the question of the ultimate limit of how many human beings could be alive at the same time. The habitable area inside the Dyson sphere could be, conservatively, at least 100 million times larger than that of Earth, maybe half a billion times larger. So, if we were to populate it with humans at a density comparable to that of the early industrial age, about 100 quadrillion (thousand trillion) humans could live there. Maybe this is what God meant when he told Adam and Eve, "Be fruitful and increase in number." But this is more than a little extreme, so we'd better explore more reasonable scenarios.

In practice, a moderate population decline may be the most likely scenario. On the positive side, it would help reduce the pressure on the environment due to industrial and agricultural activity. It would also help us to maintain a certain flow of resources into the economy, presently being depleted at an impressively fast rate. It will also reduce the need for wars, often created by expanding populations in search of vital space. It is not a bad future, but it will not be a paradise. With a declining population, we need to accept that the economy is also likely to decline. That's not a catastrophe, and cases such as Japan show that a prosperous economy can survive a population decline of about 1% per year. The use of robotics and AI may reduce the need for human power in many areas, so that it will be possible to keep the system humming along even with a smaller participation of human beings in productive tasks. But the damage already done to Earth's ecosystem is large enough that it will take a long time for it to be reabsorbed by natural processes. Its most evident manifestation is global warming. Unfortunately, even rather extreme hypotheses implying a rapid numeric decline of the human population would not necessarily reverse the warming trend.

So, what does the future have in store for humankind? From the extremes of extinction to expansion into space, there are plenty of intermediate scenarios, from the not-so-bad to the pretty good. The beauty of the future is that it is always full of possibilities.

7. Conclusion: What can you do?

The angel Gabriel was sent from God to a town of Galilee called Nazareth, to a virgin betrothed to a man named Joseph, of the house of David, and the virgin's name was Mary. And coming to her, he said, "Hail, favored one! The Lord is with you. Do not be afraid, Mary, for you have found favor with God. Behold, you will conceive in your womb and bear a son – Luke 1:26-28

Figure 42 -Raphael's 'Madonna della Seggiola' (1513-1514), Galleria degli Uffizi, Firenze. Public Domain.

It is a tradition for books dealing with potentially catastrophic futures to end with suggestions for practical action. Yet, this attitude is locked in a common way of thinking that automatically assumes that for every problem there is a solution. This idea is deeply embedded in the political discourse, with politicians forever chasing solutions to the problem of the moment. Somehow, though, despite solutions being continuously thrown at problems, the problems never seem to go away; they return, larger, nastier, and more dangerous than before. It is a rigid way of thinking that aims at maintaining things unchanged, while everything changes. In war, it leads you to more war. In commerce, it leads to bankruptcy. In love, to hate.

So, when discussing population trends, we start by trying to identify what the problem is. And we immediately bog down in a morass where what some people see as problems, others see as solutions. Are we too many? Are we becoming too few? Is more always better? What's the ideal world population? What space should we leave to the other species inhabiting planet Earth? And even before having decided what the problem is, we start throwing solutions at it.

But from what we saw in this book, the attempts by governments to intervene in population matters were often ineffective, unnecessary, or inhumane. The framing of population as a problem that must be solved is itself a trap. It's an engineer's or politician's reflex: every dial must be adjustable; every lever needs to have a predictable effect. But fertility is not a dial, it is not a lever. It's an emergent outcome of millions of individual decisions shaped by culture, economics, technology, religion, and by the very fact of being human. Central planners have been trying to turn the dial or push the levers in one or another direction. We saw pronatalism in Italy, Germany, Romania, and Hungary; anti-natalist campaigns in China, India, and other Asian countries. The scorecard is almost always disastrous: temporary blips at best, horrific human cost at worst, and in nearly every case, the trend eventually reasserted itself in the opposite direction once the effort ended.

If you are honest when you look at certain problems, you'll see that they have no solution in the common sense given to the term. That is, there is no way that governments can force natality to go the way they want it to go, and even if they could, people would disagree on which direction it should go. At the same time, there is little we can do by making individual decisions. If humans go extinct, it won't be because you decided not to have children. Nor, if they will survive, will it be because you decided to have them.

We can't engineer population trends, but we can stop pretending we could or we should. Humankind has been around for at least 300 thousand years. Our ancestors knew how to survive and how to play the reproduction game in terms of how many children to have and when. They made the right choices, and that's why we are here. Our ancestral heritage is a powerful force, and maybe, at some internal level of our minds, we understand that we have overshot the limits of this planet, and we are making a collective choice to reverse our trajectory before it is too late. Of course, we have no proof of that, and probably we never will. But we cannot, and we should not force people to make choices they don't want to make. We can focus on making whatever population trend emerges in the future more humane, ensuring people have genuine choice, reducing suffering, creating prosperity, and building resilient systems.

Nevertheless, no species lasts forever, and why should humans be an exception? Maybe we'll go extinct after having ruined the ecosystem. In that case, Earth will have to wait for a few million years to recover by itself and, perhaps, generate new sentient species who may do better than we did. Or maybe we'll find a way to stop wrecking everything around us and turn Earth into a garden. That may be what the Bible meant when it told Adam and Eve to "be fertile and multiply." Maybe we even have a chance to expand to space, and, if not us, our silicon children, the robots, will be able to do that.

We know that the future will always surprise us, and so, our job isn't to control it but to be less destructive and more adaptive as it unfolds. We need a different attitude, a way of thinking that doesn't lead us to escalation, that makes us accept the unavoidable and move along with it. In war, this attitude leads to peace; in commerce, to prosperity; in love, to love. This book is an invitation to see more clearly, resist false solutions, and participate more effectively in the open-ended story we're all still writing.

8. Acknowledgements

No part of this book was written by artificial intelligence systems, but, with the disappearance of human assistant writers, Grok 4 and Claude turned out to be good as replacements, although always with limits and a horrible tendency to hallucinate. Other AI tools I used for this book were Manus for some of the graphs, Grammarly for grammar and syntax checking, and, occasionally, DeepL for translations. Miguel Martinez (a human being) provided the final editing and smoothing of the book.

Among other human beings, I would thank the members of the Club of Rome who encouraged me to embark on this enterprise, in particular the Secretary General Carlos Álvarez Pereira, the Communication Manager Philippa Baumgartner, and the Co-presidents Paul Shrivastava and Silvia Zimmerman. I would also like to thank Chandran Nair, another Club member, for his critical review of the chapter on China. Special thanks to Dennis Meadows for having made a critical evaluation of the book and suggested the present title. I would also cite David Samways as a source of inspiration for his work with the *Journal of Population and Sustainability*. I'd like to thank my coworkers for many scientific projects in which we have been engaged together, among them Stefano Caporali, Toufic El Asmar, Sara Falsini, Alessandro Lavacchi, Gianluca Martelloni, and Ilaria Perissi. I would also like to thank the Consorzio Nazionale for the Science and Technology of Materials (INSTM) for supporting my work. The ideas of this book were inspired by many people who acted as guides for me, among them Colin Campbell, Jean Laherrere, Jorgen Randers, Matthew Simmons, and many others.

9. Appendix: Mathematical Models of Population Trends

And it came to pass in those days that a decree went out from Caesar Augustus that all the world should be registered. This census first took place while Quirinius was governing Syria. Luke 2:1-2.

Figure 42 - The Triumph of Death (Prado, Madrid), by the Flemish painter Peter Bruegel the Elder (1525-1569). Public domain. This image is communicating the severity of the plague, but it does not provide quantitative data about it. Public Domain

The appendix is dedicated to how population trends can be understood and modeled. It starts with the first attempts to measure population by means of censuses, already in use in ancient times, and then it moves on to the modern population models based on sophisticated computer-based calculations.

Visualizing Population Growth and Decline

The first step to solving a problem is to understand that it exists, and the way to do that implies the capability of collecting data about it and visualizing its trends. "Population" is an abstract concept, not directly visible to any single person. It needs to be quantified by a laborious counting process called "census" (from a Latin word meaning "to assess"). Obviously, tribal societies didn't have this capability, but the need for censuses emerged with agrarian societies, where counting the population was necessary for tax and military purposes. The Ancient Romans of imperial times performed a census every five years, an interval that gives us some idea of how expensive and time-consuming it was. One of these surveys was enacted by Emperor Tiberius around 0 CE, and it is mentioned in Luke's gospel. Censuses have been a feature of human society up to our times, part of the general tendency of governments to measure and classify everything.

The ancients used censuses to evaluate their current population, but their comments on trends of growth and decline were episodic. The Greek historian Polybius (ca. 200–118 BCE) offers a rare, direct reflection on population decline in his Histories. In Book 36, he observes that cities were emptying, fields were untended, and families had fewer children. He was probably observing the result of the agricultural collapse that occurred as a result of the overexploitation of the Greek land. In China, the scholar Ban Gu, author of the Book of Han (ca. 111 CE), records population figures from Han censuses—59 million in 2 CE, then a crash to 21 million by 57 CE after rebellions and floods. He didn't explicitly analyze "growth" or "decline" as concepts, but his matter-of-fact reporting implies rulers tracked these shifts to assess the empire's health. In Rome, the historian Livy (*Ab Urbe Condita*, 27 BCE–9 CE) described how the number of Roman citizens dropped because of the military defeats that Rome suffered against Carthage.

During the Middle Ages in Europe, there was no central authority that could carry out a census of the population. People could only understand that an epidemic was taking its toll by seeing people dying around them. That was sufficient evidence to perform appropriate rituals, or maybe to vent people's anger against whatever ethnic minority was used as a scapegoat at that time. There was no way to plot the progress of the epidemic on a graph, nor to extrapolate the trend to the future. But dramatic images, such as the "Triumph of Death" subject, served to carry the message that something was badly wrong.

In modern times, data about population growth became available, but there remained the need for a way to visualize how the population was growing or shrinking with time. That became possible with the development of graphs. For us, graphs are a normal feature of what we read and what we see on screens, but they are a modern invention. René Descartes (1596 – 1650), also known by his Latin name Cartesius, is credited with the development of the Cartesian coordinate system, the basis of "Cartesian diagrams" or "Cartesian graphs." It is said that he was inspired by watching a fly walking on a ceiling, imagining that its coordinates could be described using numbers. It is probably an apocryphal story, but in his 1637 work, La Géométrie (part of Discours de la Méthode), Descartes introduced the coordinate system that today takes his name. But it doesn't seem that he ever used what we now call a "Cartesian Graph;" these graphs popped up only in the 19th century.

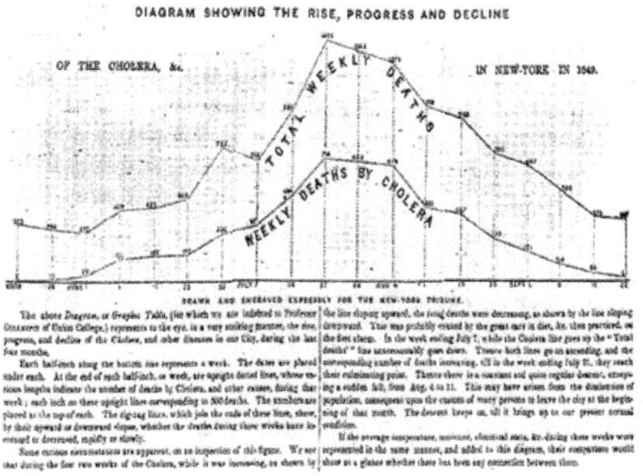

Figure 43 - Graph of the progress of the Cholera epidemic in New York up to 1849. Published in the New York Daily Tribune, it was probably the first ever published Cartesian graph.

Perhaps the first modern Cartesian Graph to be publicly available was printed in the New York Daily Tribune in 1849, which was a population graph, although dedicated to a bacterial population rather than a human one.[104]

The graph reported the progress of a cholera epidemic at the time. It was an innovation that most readers must have found challenging to understand, and, indeed, note how the editors made an effort to explain to readers the meaning of what they were seeing. From the discussions about the pandemics of 2020, it appears that even in modern times, most people were still not familiar with this kind of graph. But for anyone who has more than a passing interest in these matters, graphs are a fundamental tool. It took time before modern population graphs appeared in print. Here is the oldest example I could find: it is in the "U.S. population growth from 1790 to 1890 in the Statistical Atlas of the Eleventh Census" by Henry Gannett.[105]

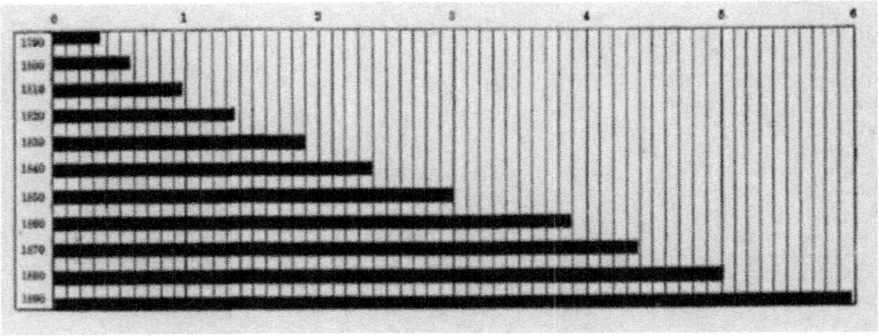

Figure 44 -- Possibly the first Cartesian graph ever published to illustrate the growth of a human population. From the "Statistical Atlas of the Eleventh Census" by Henry Gannett.

It is a Cartesian graph, but a little quirky if seen from our modern viewpoint. We tend to place the time on the x-axis, growing toward the right. Here, instead, the time goes *down* on the Y axis. But if you imagine turning the image 90 degrees anti-clockwise, the graph appears in a way that we can readily understand: a typical bar graph showing the growth trends of some quantity. Note how New York's population appears to grow exponentially, a typical behavior of a population during its early stages of expansion.

Cartesian plots of time series allow us to visualize the trends of population growth (or de-growth) in a way that tables or lists of numbers never could. Graphs make the population "alive" on paper; its history and trends bounce up to our eyes, clear and unmistakable. Over time, Cartesian graphs have evolved to describe a variety of phenomena, but their use to show the evolution of some entity as a time series remains probably the most common one.

Simple Models of Population Trends

To begin, when we refer to "population," we typically mean human or animal populations; however, nothing prevents us from using the term for non-human and non-biological entities, cats, dogs, cars, planes, etc. That said, all populations tend to follow similar laws and often predictable patterns. They may grow, remain stable, or decline.

Figure 45 – Ugo Bardi explaining the various population growth modes to his class. Note "Amelia the Amoeba," a character used to enliven the presentation. Photo by the author.

Growth. The simplest case is that of a population that expands in the environment that supports it. Typically, it doesn't just grow; it grows faster and faster: more people (or rabbits, or flies, or whatever) mean more hatchlings, and more hatchlings mean more people. Of course, no real-world population is known to have followed this growth mode forever, but this mechanism does occur for limited times.

Stability. The physical limitations of the environment dictate the limits of a population: resources or predation by another species. A population reaching this limit may attain stability, although it will often oscillate around it. Sometimes, people call that level an "attractor," and some attractors are said to be "strange" (as per the first Jurassic Park movie). It means that the oscillations are chaotic and never exactly repeat the same pattern. But that changes little to the behavior of the population.

Decline. The population may also decline or even collapse after having overcome its limits, having reached a state of "overshoot." That is, it has exploited its resources beyond their capability to reform. Often, the decline is much faster than growth as a result of additional constraints, typically pollution; in this case, one speaks of the "Seneca Effect."

Now, let's examine the details of how these mechanisms work.

Exponential (or "geometric") growth. The idea that populations tend to expand geometrically was probably known for a long time. Today, the concept of exponential economic growth is embedded in the current ways of economic thinking, but it is also common in biology. .

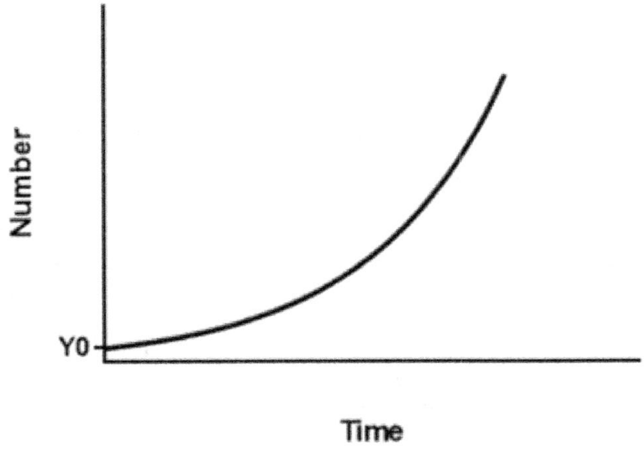

Figure 46 -- Exponential Growth

You may describe this kind of growth as in the common saying that some creatures reproduce "like rabbits." Put a couple of rabbits on a grass-covered island, and you'll soon see young rabbits appearing. These young rabbits will spawn new young rabbits, and the more rabbits there are, the faster the population will grow, at least initially. It is not just a theory; it is what happened in Australia when someone brought a few rabbits from England for the pleasure of hunting them. Disaster ensued, with rabbits growing to invade the whole of Australia, no matter how diligently human hunters tried (and still try) to exterminate them. The concept that growth generates itself is known as "positive" or "enhancing" feedback in the field of system dynamics. In chemistry, you call this behavior "autocatalytic," and in the financial world, you call it "compound interest." In mathematical terms, people speak of "geometric growth" and "exponential growth."

These are two similar, but not identical, concepts. Geometric growth nicely describes the growth of a population of, say, bacteria, that doubles at each cycle of reproduction by cellular divisions. You start with one, then you have two, then four, then eight, then 16, and so on. This is an ideal model that doesn't often occur in nature, and, in any case, it doesn't last for long. A real-world example is that of elephants in the Kruger National Park in South Africa. .

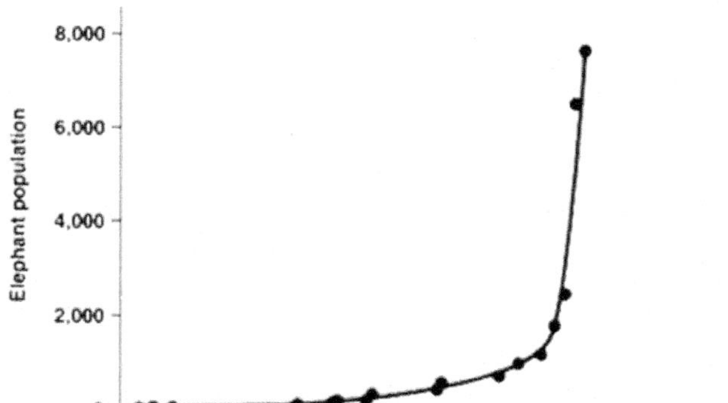

Figure 47 – Elephant population in the Kruger National Park in South Africa. Data from https://www.bio1152.nicerweb.com/

Until 1905, the park didn't host elephants, but that year, a small troop of 10 was observed, having likely moved in from a nearby region. The number of elephants rose rapidly, following a nearly exponential curve, and incidentally, demonstrating that even large animals, and not just bacteria, can show this growth pattern

Logistic growth. This idea was introduced in 1838 by the Belgian mathematician Pierre Verhulst (1804 – 1849). Verhulst wanted to give a mathematical form to Malthus' idea that the growth of a population cannot continue forever. So, he added a constraint to the equation describing exponential growth: a negative term proportional to the square of the size of the population. Since the square of the population grows faster than the population itself, this term causes growth to slow down and eventually stop. Today, it is called the "carrying capacity" of the system.

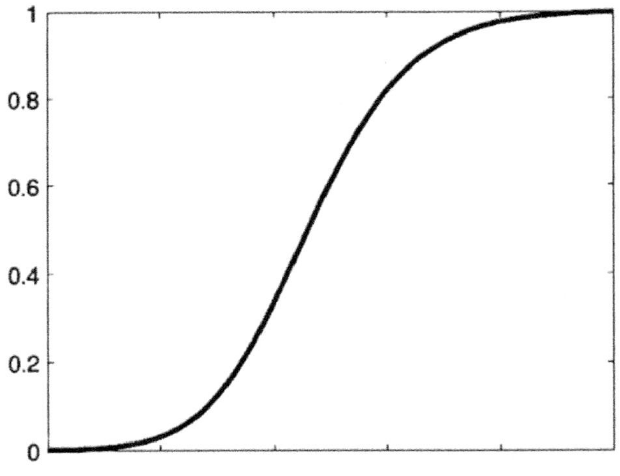

Figure 48 – the Logistic or Verhulst curve

On a Cartesian graph, the result is an "S-shaped" curve, now known as "logistic," for reasons that have to do with "skill in calculations." The idea is that the S-shaped curve provides a mathematical frame that fits with Malthus's idea that populations cannot keep growing forever. The problem with the logistic equation is that it is rare to see real-world populations that follow this growth mechanism, at least in a natural environment. To cite a case, we can go back to the example of the elephants in the Kruger National Park in South Africa. After reaching a number that appeared excessive to humans, elephants began to be culled in order to keep their numbers within a limit that seemed appropriate. The result was a growth curve that could be interpreted as a logistic. However, it was not a natural behavior; it was the result of an external intervention by human beings. In recent times, culling has been abandoned, and elephants have restarted growing again in a curve that, so far, looks exponential. It does not appear that elephants can reduce their growth rate to match the sustainability limits imposed by their environment, and that may prefigure an uncertain destiny for them.

The idea of the carrying capacity is that the growing creatures find that life becomes more and more difficult as they approach the limits of their environment to sustain their life. Herbivores have a limited area of grass to graze on, while carnivores have a limited number of herbivores to prey on. So, short of food, they reproduce more and more slowly until they reach a level where the newborn creatures replace those that die. At this point, the population becomes stable. The stability can also be generated by a predator that "culls" the population at a rate proportional to its size.

Bell-shaped curves. Malthus was an optimist when he thought that a population would reach an upper limit and then stay there. Things are not so simple in the real world, and consuming the resources that make people (or animals) live cannot be without negative consequences. In practice, populations don't just stabilize when they reach the limits of the resources that sustain them. They tend to destroy them, sometimes forever, and the result is not stabilization but collapse. In several cases, the population curve follows a "bell-shaped" curve.

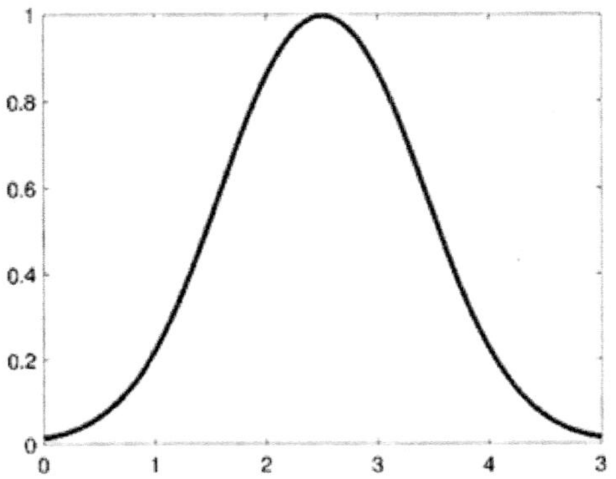

Figure 49 - A typical, bell-shaped curve The size of the population is on the Y-axis; time is on the X-axis

There is a whole fauna of mathematical functions that produce this kind of curve: the Gaussian, the derivative of a logistic, the Gompertz curve, the Dirac curve, and several others. All these curves can be used to describe the growth and decline of a biological population that runs out of the resources that make it live. It is typical of some laboratory systems, such as microorganisms grown in Petri dishes. We saw earlier the graph of the Cholera epidemic in New York in 1849. In the figure, you see another example: a curve drawn in the 1850s to illustrate the progress of the cholera pandemic in London. Pathogens can be seen as predators that live on the human population. But there is just a limited number of humans to exploit, and they gradually either die or become immune to the pathogen. When the virus, or the bacterium, has no more "food" to grow on, its population declines and eventually disappears.

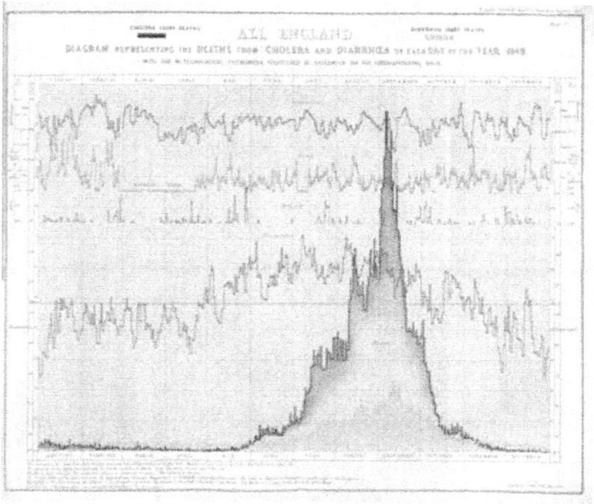

Figure 50 - Time chart of the Cholera epidemic in London in 1849. The blue line indicates the number of deaths attributed to cholera. Wikimedia Creative Commons. https://wellcomecollection.org/

The empirical observation that microorganism populations tend to follow these curves led to the development of a mathematical model to describe their behavior. In the 1920s, three British researchers, Kermack, McKendrick, and Walker, proposed the "SIR" model for the evolution of epidemics. SIR stands for "Susceptible," "Infected," and "Recovered."[106]

These models are also called "compartmental," because they divide the population into compartments. In the jargon of these studies, the number of individuals belonging to each of these categories is referred to as a "stock." Note also that the model is based on data on the effects of the pathogen, not directly on the pathogen population, which is usually impossible to measure. Instead, the number of infected people is a proxy for the number of microorganisms. The idea of the SIR model is that a population starts with a small fraction of its members infected. As more people get infected, the number of susceptible individuals declines while that of infected ones increases. As time goes by, infected people recover and become "immune" (or die). Eventually, the pathogens don't find enough targets to continue spreading, and their population (counted in terms of the number of infected people) starts declining, too. At that point, the human population is said to have reached "herd immunity," and the epidemic doesn't grow anymore. In time, the epidemic abates, and the microorganism population goes down to zero. Epidemiological models exist in several variants, and there are plenty of bells and whistles that can be added to simple models to make them complicated (and often not more reliable). In some cases, these additional elements can be used to torture the data until it confesses to whatever an investigator wants it to confess. But that's true for all models.

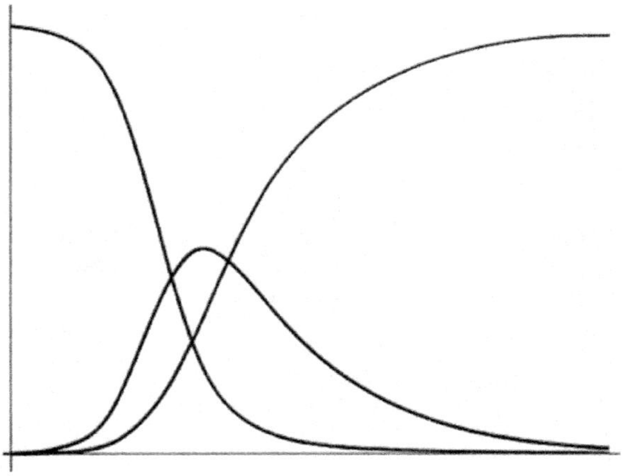

Figure 51 - Typical results of the SIR model. The time is on the x-axis, while the three curves describe, respectively, the number of susceptible, infected, and recovered people107

Nothing prevents the SIR model from being applied to populations other than microorganisms. In practice, however, it remains traditionally confined to medical studies. In the more general field of population biology, similar models are used, but with different names. The ancestor of these models was independently developed a few years before the SIR model by Alfred Lotka (a physical chemist) and Vito Volterra (a mathematician). Today, the model takes the name of the "Lotka-Volterra" or "LV" model, sometimes the "prey/predator" or "Rabbits/Foxes" model. In its simplest version, the Lotka-Volterra model is nearly identical to the SIR model, with the "susceptible" stock called "prey," and the "infected" stock called the "predator." The only difference is that in the LV model, the prey population reproduces and can grow, while in the SIR model, it does not. There are several variants to the LV model that can become much more complicated on the basis of different assumptions. [52]

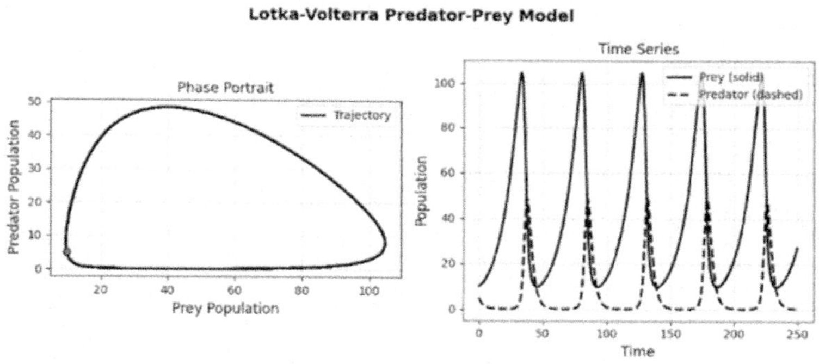

Figure 52 - Typical results of the Lotka-Volterra model. Calculated using the standard LV equations

In the LV model, the predators never succeed at completely exterminating their prey. But they can reduce the number of preys to a level so low that it can't support so many predators. At this point, predators start dying, and the prey population starts re-growing. The result of the simplest version of the LV model is a series of oscillations that extends to infinity. The "phase portrait" graph plots the number of predators as a function of the number of the prey stock. Note how the two populations chase each other without ever reaching a stable point. If the model is modified by adding a "friction" term, then the predator and prey stock will eventually reach a condition of stability. In the natural world, populations seem to be able to reach that condition, unless they are influenced by external perturbations.

The Lotka-Volterra model becomes the same as the SIR model if you remove the term that generates the growth of the prey. In this case, it is sometimes called the "Single-cycle Lotka-Volterra" (SCLV) model[107]. This model is related to the one called the "Hubbert" model, which was developed outside population studies but turned out to be part of the family. Marion King Hubbert (1903-1989) was a US petroleum geologist who developed a model on the oil extraction cycle in the 1950s.

The Seneca Curve.

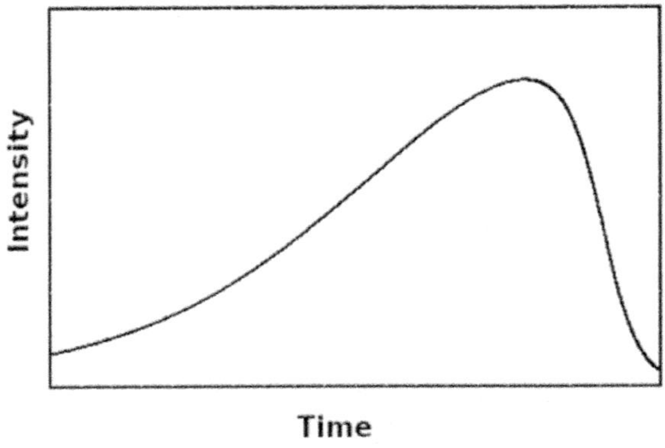

Figure 53 - The Seneca curve. Growth is slow, but collapse is rapid. Calculated according to the model described in Ugo Bardi's Book The Seneca Effect (2017)

The "Seneca Curve" or "Seneca Effect" are terms derived from a statement by the Roman Philosopher Lucius Annaeus Seneca, who, in one of his letters, noted how "Growth is slow, but ruin is rapid." It is a concept that describes well how some populations tend to crash rather than just decline.

The Seneca Model can be described mathematically by a set of three differential equations. In simpler terms, it is the result of a "perfect storm" that affects a growing population. If the decline is caused by a single factor, such as running out of food, then the curve tends to be the symmetric, bell-shaped one described in the previous section. However, if additional factors appear together to weigh on the population, then decline is typically faster than growth. An example can be the role of predators, added to the lack of food. In the case of a human population, famine may be coupled to a plague or, in recent times, to pollution

.

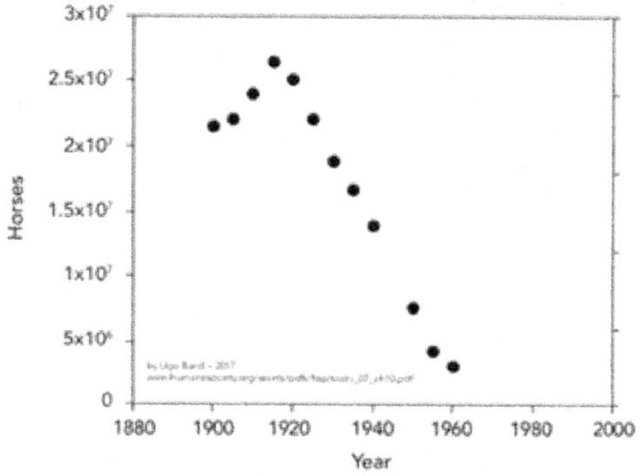

Figure 54 - The Horse population in the United States. Data from www.thehumanesociety.org (no longer available).

A typical example of a real-world population collapse that followed a Seneca trajectory is that of the horse population in the United States, which rapidly declined after horses were no longer used as a means of transportation. Later, the horse population restarted growing, but it remains today much smaller than it was at its peak.

The "Hokusai Model" – Using this term, I am referring to the famous print of "The Wave" by the Japanese painter Katsushika Hokusai

Figure 55 - Katsushika Hokusai, 'The Great Wave off Kanagawa' (c. 1830-1833), from the series 'Thirty-Six Views of Mount Fuji.

It is a visual help to understand that when things go bad, they gang together to do much worse things. As Shakespeare said in "Hamlet," troubles come "not as single spies, but in battalions." A population collapse may come in the shape of a Seneca Curve, but even though the curve shows a sharp downward trajectory, it is still a smooth curve. The "Hokusai model" refers to a situation in which the system goes chaotic or unpredictable. It refers to the various forms of rapid collapse, which are sometimes referred to as "Black Swans" [108] or "Dragon Kings,"[109] part of the general field of "critical" phenomena. [110] These are real phenomena that happen in the real world. Examples are earthquakes, avalanches, landslides, tsunamis, and more.

The models described here are at the basis of population studies in biology and medicine. They are also often used to describe the behavior of human populations, but they form a separate and different compartment of science from the main field of human population study: demography. It is a field strongly influenced by economics, and uses an approach mainly based on curve fitting. Demographers tend to try to learn about population trends starting from historical data, rather than supposing that the curve will follow a specific curve or theoretical model.

Demographic studies use sophisticated statistical methods to analyze data, identify trends, and project them into the future. "Curve Fitting" is a process in which the experimental or historical data are compared to a curve obtained by a model or an equation. The correspondence will never be perfect, and there exist sophisticated statistical treatments that can quantify the discrepancy between theory and the real world. These methods can be used to decide whether a specific model is good or not to describe a process, for instance, the spreading of an epidemic. Once a model has been tested and shown to produce a good fit, it can be extrapolated to obtain a peek at the future. It is an ensemble of methods that goes under the name of "MLE," "Maximum Likelihood Estimation." Sometimes, these routines are based on the basic population models described here, or it is possible to ask the calculating routines to estimate the equation or set of equations that's behind the historical trends. Then, the system can extrapolate the trend into the future. As you may easily understand, this approach is fraught with uncertainties.

The model used by the United Nations for the World Population Prospects (WPP) is one of the most advanced. It relies on a variety of data sources to build its models, such as National Censuses to provide baseline population counts, typically conducted every 5–10 years by countries. Birth and death registrations from national systems are also used, and special surveys are often carried out. For countries with limited or unreliable data, the UN uses statistical methods to estimate missing figures based on historical trends or data from similar regions.

The model is based on a cohort-component method, a standard demographic technique, to project population changes over time. It tracks population by age and sex, factoring in three key drivers: 1) the Total Fertility Rate (TFR), or average number of births per woman, 2) Mortality, and 3) Migrations. Assumptions about future fertility, mortality, and migration are based on trends in education, healthcare, economic development, and policy. Each "cohort" (e.g., women aged 20–24) moves through time, aging and experiencing these demographic processes, which are calculated in five-year intervals.

Since the future is uncertain, the UN uses a Bayesian hierarchical model to generate probabilistic projections (introduced in the 2015 WPP and refined since), incorporating uncertainty by producing thousands of possible trajectories for TFR and life expectancy. The output is a median projection (most likely scenario) alongside ranges (e.g., 80% or 95% confidence intervals), reflecting high and low variants.

The UN model, sophisticated as it is, does not include factors such as climate change, effects of pollution, and resource depletion. They are supposed to be incorporated in the past trends, and hence to be "virtually" accounted for in the future. There is growing pressure to better integrate climate change into demographic modeling through interdisciplinary models that merge climate, economic, and population dynamics. However, the problem is that including more details adds uncertainty to the calculations and makes the models less robust.

One problem is that since historical data do not usually involve population crashes, the method will never predict a possible future crash. Predicting the future is always difficult, and no matter how good your model is, it will never be the same as the real world. There will always be something in the real world that its virtual counterpart in the memory of a computer didn't take into account, but which turned out to be important. Nevertheless, we all know that life is fraught with uncertainties, and often, scientific forecasters can't do much better than the average guy in the street. Otherwise, they would be richer and more successful. So, we keep going, trying to do our best.

Dynamic Models: The Limits to Growth

From Malthus onward, modeling population trends was a question of using equations that could be written on paper and solved by hand. That was often laborious and time-consuming. The revolution came with the development of the first digital computers in the 1950s and 1960s. At that point, the drudgery of finding approximate solutions to a set of equations could be left to the machine, while humans could concentrate on the structure of the model. This kind of modeling was initially developed for military applications, but it was very versatile and could be dedicated to modeling population growth and decline.

One of the pioneers in this field was a professor at the Massachusetts Institute of Technology in the 1960s, Jay Forrester (1918-2016), who built his own advanced computers to run his sophisticated models. Forrester developed a new field of modeling called "System Dynamics," which used a specific programming language called DYNAMO. Already in the 1960s, Forrester thought that he could use his computers and his models to study the evolution of the world's economic system, including the trajectory of the human population.

In 1969, Forrester was a faculty member of MIT when he was contacted by Aurelio Peccei, the president of the recently formed Club of Rome. At that time, Peccei was interested in understanding how Earth's resources could be distributed among the people of the world. But no data or models were available to understand how exactly the economic system would evolve in a situation that, at that time, seemed to promise a future of abundance. Would there be enough for a fair share for everyone? But how would the production of goods adapt to a growing population? The existing models were either partial or oversimplified.

Forrester's calculation methods promised to generate answers to these questions. After discussing with Forrester, Peccei found a sponsor for the project with the Volkswagen Foundation, which provided a grant. That led to the creation of the first "world models," which examined the whole Earth's economic system, taking into account the physical constraints of the environment.

The first study was created by Forrester himself and published in 1971 with the title *World Dynamics* [9]. A second study, published in 1972 by a group of Forrester's coworkers, carried the title The Limits to Growth (LTG)[10]. These studies were milestones in the understanding of the evolution of the human population as a function of the economy. The results were similar and showed that Malthus was an optimist.

Far from reaching the limits to growth and staying there, human civilization was to overshoot the limits and then crash down. The figure shows the results of the "business as usual" (or "base case") scenario

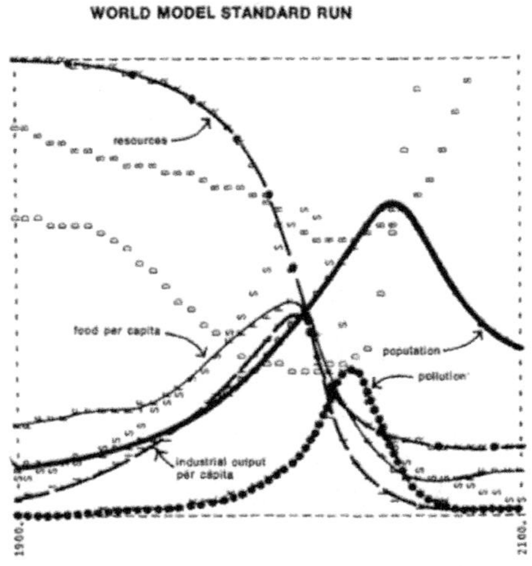

Figure 56 - The "base case" scenario of the 1972 edition of The Limits to Growth. Image courtesy of Dennis Meadows

. You can see in the figure the original graph that was printed in the first edition of LTG, plotted in black and white with the limited plotting capabilities of the time. It is the so-called "base case" scenario, a projection of how the world's social and economic system would evolve if the trends of the time were to be maintained. It was based on the best available data on the quantities of mineral resources, land, and industrial capital. The results for the global population are shown as the black curve, the one that reaches the highest level in the mid-21st century and then rapidly declines.

It was a revolutionary result. It was the first time that a population was quantitatively studied not as an independent variable based on statistical trends, but as part of a complex system that dynamically changes over time. Not that people hadn't realized that the population would depend on such factors as the availability of food and energy, but before the development of system dynamics, it wasn't impossible to calculate all these factors together.

Let's describe more in detail how the "world modeling" approach worked. In the case of The Limits to Growth, its first appearance in 1972 was based on a mathematical model called "World3." It had the "3" index because it was the third of a series that had been started earlier on by Forrester, who had used the "World2" model for the study he published in 1971. The World3 model was remarkably complex for its time. It had hundreds of parameters arranged in a complicated network of interactions. Nevertheless, the model was based on five main parameters:

- Human Population
- Non-renewable Resources (minerals)
- Renewable Resources (agriculture)
- Capital Resources
- Pollution

These five elements, also called "stocks," interact with each other by exchanging materials, energy, and money. The main results of the model could be explained in qualitative terms by describing how the stocks evolved together. The combined effects of resource depletion and pollution accumulation gradually reduced the ability of the industrial system to accumulate capital, which was created by the resources and had to be spent to fight pollution. With less capital available, both the industrial system and the agricultural one slowed down their growth and, eventually, reached a peak and then declined. Later, the population stock would follow the same trajectory, as impoverished families would be unable to form families and have children. With the collapse, pollution, too, would decline since it was created mainly by the industrial system.

These results were not to be taken as predictions but just as answers to the question "what if?" They formed a series of scenarios that demonstrated the robustness of the tendency toward collapse. Unless some specific interventions were to be carried out, or some miraculous new technology developed, a global collapse was unavoidable at some moment during the 21st century, probably during the first half of it. Only specific policies to curb growth could avoid it. In general, the "base-case" scenario, the one that assumed no changes in the current policies at that time, produced results that turned out to describe the world's situation as it is today. Re-examinations of the study, such as the one by Graham Turner in 2008[111] and by Gaia Herrington in 2021[112] showed that the correspondence with the historical trends was not perfect, but remarkably good considering that it spanned more than half a century.

The population parameter in *The Limits to Growth* was one of the most uncertain and most difficult to deal with. Most of the parameters of the model could be treated in terms of the standard assumptions of economics. That is, it could be assumed that people would operate in such a way to maximize their economic benefit in the short run. In more mundane terms, you would say that they were trying to make as much money as they could. But population growth (or decline) depends on choices that cannot be based on monetary gains, but on the more vaporous concept of "utility." That implies proposing what factors make people decide to have children or not. It is a decision that most likely results from a series of factors combining: food availability, cost of living, future prospects for one's children, local customs and morals, and more. The authors of the first Limits to Growth study used the known historical data on the relationship between wealth and birthrate. This relation is far from being linear: poor people tend to have more children when they get a little richer. But, over a certain limit, the "population transition" kicks in, and birthrates decline. It is a well-known phenomenon, normally ascribed to social factors.

If we assume that the correlation remains valid when economic growth becomes a decline, then one would expect that people would have more children. That explains why the Limits to Growth model shows population growing while everything else collapses. But that's hardly a realistic assumption: hungry and sick people are unlikely to think very much of sex. Besides, the assumption doesn't correspond to the data we have about societies that actually collapsed in recent times, such as the states that once formed the Soviet Union. In all recent cases, the population went down rapidly, together with the economy.

This problem was noted by the authors of the 1972 study, and it was corrected with more realistic assumptions that placed more importance on wealth and resources on people's willingness to have children. In the new versions of the study, the decline of the world's population starts about one decade after the start of the decline of food availability. Other studies based on system dynamics provided similar results.

Most dynamic models generate the turning point at some moment during the first half of the 21st century. It is a prediction that still needs to be verified, but it looks reasonable considering that already today, one-third of the world's population lives in countries where the population is declining. As always, the future is not exactly predictable, but these models provide us with tools to be prepared for what will befall us.

10. References

(1) Láng-Ritter, J.; Keskinen, M.; Tenkanen, H. Global Gridded Population Datasets Systematically Underrepresent Rural Population. *Nat. Commun.* **2025**, *16* (1), 2170. https://doi.org/10.1038/s41467-025-56906-7.

(2) Greenspoon, L.; Krieger, E.; Sender, R.; Rosenberg, Y.; Bar-On, Y. M.; Moran, U.; Antman, T.; Meiri, S.; Roll, U.; Noor, E.; Milo, R. The Global Biomass of Wild Mammals. *Proc. Natl. Acad. Sci.* **2023**, *120* (10), e2204892120. https://doi.org/10.1073/pnas.2204892120.

(3) Rockström, J.; Steffen, W.; Noone, K.; Persson, Å.; Chapin, F. S.; Lambin, E. F.; Lenton, T. M.; Scheffer, M.; Folke, C.; Schellnhuber, H. J.; Nykvist, B.; de Wit, C. A.; Hughes, T.; van der Leeuw, S.; Rodhe, H.; Sörlin, S.; Snyder, P. K.; Costanza, R.; Svedin, U.; Falkenmark, M.; Karlberg, L.; Corell, R. W.; Fabry, V. J.; Hansen, J.; Walker, B.; Liverman, D.; Richardson, K.; Crutzen, P.; Foley, J. A. A Safe Operating Space for Humanity. *Nature* **2009**, *461* (7263), 472–475. https://doi.org/10.1038/461472a.

(4) *Earth beyond six of nine planetary boundaries*. https://doi.org/10.1126/sciadv.adh2458.

(5) United Nations. *World Population Prospects*. United Nations. https://population.un.org/wpp/ (accessed 2025-04-07).

(6) Dhakad S. et al. *Updating the Shared Socioeconomic Pathways (SSPs) Global Population and Human Capital Projections.*; IIASA. http://iiasa.ac.at/models-tools-data/ssp-2023 (accessed 2025-04-07).

(7) Lee, H.; Romero, J. *Sixth Assessment Report — IPCC*; IPCC, 2024. https://www.ipcc.ch/assessment-report/ar6/ (accessed 2024-07-23).

(8) Vollset, S. E.; Goren, E.; Yuan, C.-W.; Cao, J.; Smith, A. E.; Hsiao, T.; Bisignano, C.; Azhar, G. S.; Castro, E.; Chalek, J.; Dolgert, A. J.; Frank, T.; Fukutaki, K.; Hay, S. I.; Lozano, R.; Mokdad, A. H.; Nandakumar, V.; Pierce, M.; Pletcher, M.; Robalik, T.; Steuben, K. M.; Wunrow, H. Y.; Zlavog, B. S.; Murray, C. J. L. Fertility, Mortality, Migration, and Population Scenarios for 195 Countries and Territories from 2017 to 2100: A Forecasting Analysis for the Global Burden of Disease Study. *The Lancet*

2020, *396* (10258), 1285–1306. https://doi.org/10.1016/S0140-6736(20)30677-2.
(9) Forrester, J. *World Dynamics*; Wright-Allen Press: Cambridge, Massachusetts, 1971.
(10) Meadows, D.; Meadows, D. H.; Randers, J.; Behrens, W., Jr. *The Limits to Growth*; Potomac Associates, 1972.
(11) Ugo Bardi. *The Seneca Effect. Why Growth Is Slow but Collapse Is Rapid.*; Springer Verlag, 2017.
(12) Levine, H.; Jørgensen, N.; Martino-Andrade, A.; Mendiola, J.; Weksler-Derri, D.; Mindlis, I.; Pinotti, R.; Swan, S. H. Temporal Trends in Sperm Count: A Systematic Review and Meta-Regression Analysis. *Hum. Reprod. Update* **2017**, *23* (6), 646–659. https://doi.org/10.1093/humupd/dmx022.
(13) Calhoun, J. B. Death Squared: The Explosive Growth and Demise of a Mouse Population. *Proc. R. Soc. Med.* **1973**, *66* (1 Pt 2), 80–88. https://doi.org/10.1177/00359157730661P202.
(14) Herculano-Houzel, S. *The Human Advantage: A New Understanding of How Our Brain Became Remarkable*; The MIT Press, 2016. https://doi.org/10.7551/mitpress/9780262034258.001.0001.
(15) Pinson, A.; Xing, L.; Namba, T.; Kalebic, N.; Peters, J.; Oegema, C. E.; Traikov, S.; Reppe, K.; Riesenberg, S.; Maricic, T.; Derihaci, R.; Wimberger, P.; Pääbo, S.; Huttner, W. B. Human TKTL1 Implies Greater Neurogenesis in Frontal Neocortex of Modern Humans than Neanderthals. *Science* **2022**, *377* (6611), eabl6422. https://doi.org/10.1126/science.abl6422.
(16) Bardi, U. The Role of Carbon Dioxide in the Evolution of Intelligence: Solving the Dinosauroid Question. *Qeios* **2025**. https://doi.org/10.32388/IKNUZU.
(17) Morris, D. *The Naked Ape*; Jonathan Cape, 1967.
(18) Gould, S. J.; Vrba, E. S. Exaptation-A Missing Term in the Science of Form. *Paleobiology* **1982**, *8* (1), 4–15.
(19) Bartsiokas, A.; Arsuaga, J.-L. Hibernation in Hominins from Atapuerca, Spain Half a Million Years Ago. *L'Anthropologie* **2020**, *124* (5), 102797. https://doi.org/10.1016/j.anthro.2020.102797.
(20) Roebroeks, W.; MacDonald, K.; Scherjon, F.; Bakels, C.; Kindler, L.; Nikulina, A.; Pop, E.; Gaudzinski-Windheuser, S. Landscape Modification by Last Interglacial Neanderthals. *Sci. Adv.* **2021**, *7* (51), eabj5567. https://doi.org/10.1126/sciadv.abj5567.

(21) Hawks, J. Significance of Neandertal and Denisovan Genomes in Human Evolution*. *Annu. Rev. Anthropol.* **2013**, *42* (Volume 42, 2013), 433–449. https://doi.org/10.1146/annurev-anthro-092412-155548.
(22) Bardi, U. *Chimeras: Inanna and Ebih: a report of an ancient ecological catastrophe?*. Chimeras. https://chimeramyth.blogspot.com/2015/08/inanna-and-ebih-report-of-ancient.html (accessed 2021-11-09).
(23) Bairoch, P. *Cities and Economic Development: From the Dawn of History to the Present*; University of Chicago Press: Chicago, Il, 1988.
(24) Varia. *World Population Prospects*. World Population Prospects. https://population.un.org/wpp/ (accessed 2025-09-14).
(25) Greiling, W. *Wie werden wir leben?*; Econ-Verlag, 1954.
(26) Lutz, W.; Sanderson, W.; Scherbov, S. Doubling of World Population Unlikely. *Nature* **1997**, *387* (6635), 803–805. https://doi.org/10.1038/42935.
(27) Lutz, W.; Sanderson, W.; Scherbov, S. The End of World Population Growth. In *The End of World Population Growth in the 21st Century*; Lutz, W., Sanderson, W., Scherbov, S., Eds.; Earthscan: London, 2004; pp 17–84.
(28) Nebel, A.; Kling, A.; Willamowski, R.; Schell, T. Recalibration of Limits to Growth: An Update of the World3 Model. *J. Ind. Ecol.* **2024**, *28* (1), 87–99. https://doi.org/10.1111/jiec.13442.
(29) Bystroff, C. Footprints to Singularity: A Global Population Model Explains Late 20th Century Slow-down and Predicts Peak within Ten Years. *PLoS ONE* **2021**, *16* (5), e0247214. https://doi.org/10.1371/journal.pone.0247214.
(30) Randers, J. *2052: A Global Forecast for the Next Forty Years*, Illustrated edition.; Chelsea Green: White River Junction, Vermont, 2012.
(31) Dixson-Decleve, S.; Gaffney, O.; Ghosh, J.; Randers, J.; Rockstrom, J.; Stoknes, P. E. *Earth for All: A Survival Guide for Humanity*; New Society Publishers: Gabriola Island, British Columbia, Canada, 2022.
(32) Meadows, D. H.; Meadows, D. L.; Randers, J. *Beyond the Limits*; Chelsea Green: Post Mills, Vermont, USA, 1992.
(33) Meadows, D. H.; Randers, J.; Meadows, D. L. *Limits to Growth: The 30 Year Update*; Chelsea Green: White River Junction, 2004.

(34) Hotelling, H. The Economics of Eshaustible Resources. *J. Polit. Econ.* **1931**, *39* (2), 137–175.
(35) Zimmermann, E. W. *World Resources and Industries;: A Functional Appraisal of the Availability of Agricultural and Industrial Resources,*; Harper & Brothers: New York, 1933.
(36) Daly, H. E. Georgescu-Roegen versus Solow/Stiglitz. *Ecol. Econ.* **1997**, *22* (3), 261–266. https://doi.org/10.1016/S0921-8009(97)00080-3.
(37) Hall, C.; Lindenberger, D.; Kümmel, R. The Need to Reintegrate the Natural Sciences with Economics: Neoclassical Economics, the Dominant Form of Economics Today, Has at Least Three Fundamental Flaws. *BioScience* **2001**.
(38) Hubbert, M. K. Nuclear Energy and the Fossil Fuels. In *Spring Meeting of the Southern District, American Petroleum Institute, Plaza Hotel, San Antonio, Texas, March 7–8-9,*; San Antonio, Texas, 1956.
(39) Bardi, U.; Lavacchi, A. A Simple Interpretation of Hubbert's Model of Resource Exploitation. *Energies* **2009**, *2* (3), 646–661. https://doi.org/10.3390/en20300646.
(40) Hardin, G. The Tragedy of the Commons. *Science* **1968**, *162* (13 December), 1243–1248.
(41) Ostrom, E. *Governing the Commons: The Evolution of Institutions for Collective Action*; Cambridge University Press: Cambridge, UK, 1990.
(42) Moxnes, E. Not Only the Tragedy of the Commons: Misperceptions of Feedback and Policies for Sustainable Development. *Syst. Dyn. Rev.* **2000**, *16* (4), 325–348.
(43) Gorshkov, V.G; Makarieva, A.M. Biotic Pump of Atmospheric Moisture as Driver of the Hydrological Cycle on Land. *Hydrol. Earth Syst. Sci. Discuss.* **2006**, *3*, 2621–2673.
(44) Kralj-Fišer, S.; Čelik, T.; Lokovšek, T.; Šuen, K.; Šiling, R.; Kuntner, M. Development and Growth in Synanthropic Species: Plasticity and Constraints. *Naturwissenschaften* **2014**, *101* (7), 565–575. https://doi.org/10.1007/s00114-014-1194-y.
(45) Calhoun, J. B. A Behavioral Sink. In *Roots of Behavior*; National Institutes of Health, 1962.
(46) Ramsden, E.; Adams, J. Escaping the Laboratory: The Rodent Experiments of John B. Calhoun & Their Cultural Influence. *J. Soc. Hist.* **2009**, *42* (3), 761–792.

(47) *Guide for the Care and Use of Laboratory Animals: Eighth ...* m.moam.info. https://m.moam.info/guide-for-the-care-and-use-of-laboratory-animals-eighth-_59c5d0d01723dd42ad22462f.html (accessed 2025-05-15).

(48) Terman, C. R. A Study of Population Growth and Control Exhibited in the Laboratory by Prairie Deermice. *Ecology* **1965**, *46* (6), 890–895. https://doi.org/10.2307/1934030.

(49) Suvorov, A. Population Numbers and Reproductive Health. *Endocrinology* **2021**, *162* (11), bqab154. https://doi.org/10.1210/endocr/bqab154.

(50) Dolnik, V. R. How Many People Is the Earth Designed For? *Priroda* **1992**, *6* (922), 3–16.

(51) Bonenfant, C.; Gaillard, J.; Coulson, T.; Festa-Bianchet, M.; Loison, A.; Garel, M.; Loe, L. E.; Blanchard, P.; Pettorelli, N.; Owen-Smith, N.; Du Toit, J.; Duncan, P. Chapter 5 Empirical Evidence of Density-Dependence in Populations of Large Herbivores. In *Advances in Ecological Research*; Academic Press, 2009; Vol. 41, pp 313–357. https://doi.org/10.1016/S0065-2504(09)00405-X.

(52) Tyutyunov, Y.; Titova, L. From Lotka-Volterra to Arditi-Ginzburg: 90 Years of Evolving Trophic Functions // Biology Bulletin Reviews. **2020**, *10*, 167–185. https://doi.org/10.1134/S207908642003007X.

(53) Sibly, R. M.; Barker, D.; Denham, M. C.; Hone, J.; Pagel, M. On the Regulation of Populations of Mammals, Birds, Fish, and Insects. *Science* **2005**, *309* (5734), 607–610. https://doi.org/10.1126/science.1110760.

(54) Peen, J.; Schoevers, R. A.; Beekman, A. T.; Dekker, J. The Current Status of Urban-Rural Differences in Psychiatric Disorders. *Acta Psychiatr. Scand.* **2010**, *121* (2), 84–93. https://doi.org/10.1111/j.1600-0447.2009.01438.x.

(55) Makarieva, A. *Does the concept of 'pollution' match the complexity of human–biosphere interactions?* Biotic Regulation and Biotic Pump. https://bioticregulation.substack.com/p/does-the-concept-of-pollution-match (accessed 2025-05-09).

(56) NIH. *Mental Illness - National Institute of Mental Health (NIMH).* https://www.nimh.nih.gov/health/statistics/mental-illness (accessed 2025-04-11).

(57) Ridenour, N. *Mental Health in the United States, a Fifty-Year History*; Cambridge, Mass., Commonwealth Fund, 1961.
(58) Inc, G. *Global Emotions Report*; 2024. https://www.gallup.com/analytics/349280/gallup-global-emotions-report.aspx (accessed 2025-04-03).
(59) MacLeod, J. The Metabolism of Human Spermatozoa. *Proc. Soc. Exp. Biol. Med.* **1939**, *42* (1), 153–155. https://doi.org/10.3181/00379727-42-10830.
(60) Carlsen, E.; Giwercman, A.; Keiding, N.; Skakkebaek, N. E. Evidence for Decreasing Quality of Semen during Past 50 Years. *BMJ* **1992**, *305* (6854), 609–613. https://doi.org/10.1136/bmj.305.6854.609.
(61) Wang, C.; Mbizvo, M.; Festin, M. P.; Björndahl, L.; Toskin, I. Evolution of the WHO "Semen" Processing Manual from the First (1980) to the Sixth Edition (2021). *Fertil. Steril.* **2022**, *117* (2), 237–245. https://doi.org/10.1016/j.fertnstert.2021.11.037.
(62) Rahban, R.; Priskorn, L.; Senn, A.; Stettler, E.; Galli, F.; Vargas, J.; Van den Bergh, M.; Fusconi, A.; Garlantezec, R.; Jensen, T. K.; Multigner, L.; Skakkebæk, N. E.; Germond, M.; Jørgensen, N.; Nef, S.; Group, the N. W. Semen Quality of Young Men in Switzerland: A Nationwide Cross-Sectional Population-Based Study. *Andrology* **2019**, *7* (6), 818–826. https://doi.org/10.1111/andr.12645.
(63) Fierz, L. *Kastriert im Mutterleib: Jeder ist Versuchskaninchen*; BoD – Books on Demand: Hamburg, 2025.
(64) Nihart, A. J.; Garcia, M. A.; El Hayek, E.; Liu, R.; Olewine, M.; Kingston, J. D.; Castillo, E. F.; Gullapalli, R. R.; Howard, T.; Bleske, B.; Scott, J.; Gonzalez-Estrella, J.; Gross, J. M.; Spilde, M.; Adolphi, N. L.; Gallego, D. F.; Jarrell, H. S.; Dvorscak, G.; Zuluaga-Ruiz, M. E.; West, A. B.; Campen, M. J. Bioaccumulation of Microplastics in Decedent Human Brains. *Nat. Med.* **2025**, 1–6. https://doi.org/10.1038/s41591-024-03453-1.
(65) Ponnana, S. R.; Rajagopalan, T.; Zhang, T.; Chen, Z.; Sirasapalli, S. K.; Deo, S. Microplastic Concentration, Social, and Environmental Features and Their Association with Chronic Disease Prevalence: An Analysis across u.s. Census Tracts. *JACC* **2025**, *85* (12_Supplement), 374–374. https://doi.org/10.1016/S0735-1097(25)00858-7.

(66) UNEP, U. N. *Chemicals in Plastics - A Technical Report*; UNEP, 2023. http://www.unep.org/resources/report/chemicals-plastics-technical-report (accessed 2024-06-11).

(67) Mastrantonio, M.; Bai, E.; Uccelli, R.; Cordiano, V.; Screpanti, A.; Crosignani, P. Drinking Water Contamination from Perfluoroalkyl Substances (PFAS): An Ecological Mortality Study in the Veneto Region, Italy. *Eur. J. Public Health* **2018**, *28* (1), 180–185. https://doi.org/10.1093/eurpub/ckx066.

(68) Di Nisio, A.; Sabovic, I.; Valente, U.; Tescari, S.; Rocca, M. S.; Guidolin, D.; Dall'Acqua, S.; Acquasaliente, L.; Pozzi, N.; Plebani, M.; Garolla, A.; Foresta, C. Endocrine Disruption of Androgenic Activity by Perfluoroalkyl Substances: Clinical and Experimental Evidence. *J. Clin. Endocrinol. Metab.* **2019**, *104* (4), 1259–1271. https://doi.org/10.1210/jc.2018-01855.

(69) Edaes, F. S.; de Souza, C. B. BPS and BPF Are as Carcinogenic as BPA and Are Not Viable Alternatives for Its Replacement. *Endocr. Metab. Immune Disord. Drug Targets* **2022**, *22* (9), 927–934. https://doi.org/10.2174/1871530322666220316141032.

(70) Chao, F.; Gerland, P.; Cook, A. R.; Alkema, L. Systematic Assessment of the Sex Ratio at Birth for All Countries and Estimation of National Imbalances and Regional Reference Levels. *Proc. Natl. Acad. Sci.* **2019**, *116* (19), 9303–9311. https://doi.org/10.1073/pnas.1812593116.

(71) Press, A. More than 63 Million Women "missing" in India, Statistics Show. *The Guardian*. January 30, 2018. https://www.theguardian.com/world/2018/jan/30/more-than-63-million-women-missing-in-india-statistics-show (accessed 2025-11-27).

(72) Miguel, E. Poverty and Witch Killing. *Rev. Econ. Stud.* **2005**, *72* (4), 1153–1172.

(73) United, N. *World Population Prospects as Assessed in 1963*; 41; 1966. https://www.un.org/development/desa/pd/content/world-population-prospects-assessed-1963 (accessed 2025-02-16).

(74) UCSUSA. *1992 World Scientists' Warning to Humanity"*. https://www.ucsusa.org/resources/1992-world-scientists-warning-humanity (accessed 2025-02-22).

(75) Simmons, M. *Revisiting The Limits to Growth: Could the Club of Rome have been correct, after all? An energy white paper.*

http://www.greatchange.org/ov-simmons,club_of_rome_revisted.html (accessed 2014-01-01).
(76) Gilfilan, S. C. Lead Poisoning and the Fall of Rome. *J. Occup. Environ. Med.* **1965**, *7* (1), 53–60.
(77) Cilliers, L.; Retief, F. Lead Poisoning and the Downfall of Rome: Reality or Myth? In *History of Toxicology and Environmental Health: Toxicology in Antiquity*; Wexler, P., Ed.; Elsevier: Amsterdam, Boston, Heidelberg, London, Nw York, Oxford, Paris, San Diego, San Francisco, Singapore, Sydney, Tokyo, 2014.
(78) Colás, R. Ancient Lead, Mercury, and Tin. In *Reverse Engineering of Ancient Metals*; Carrizo, P. S., Ed.; Springer International Publishing: Cham, 2021; pp 29–40. https://doi.org/10.1007/978-3-030-72842-7_3.
(79) Büntgen, U. European Climate Variability and Human Susceptibility over the Past 2500 Years. **2010**.
(80) Makarieva, A. M.; Nefiodov, A. V.; Nobre, A. D.; Baudena, M.; Bardi, U.; Sheil, D.; Saleska, S. R.; Molina, R. D.; Rammig, A. The Role of Ecosystem Transpiration in Creating Alternate Moisture Regimes by Influencing Atmospheric Moisture Convergence. *Glob. Change Biol.* **2023**, *29* (9), 2536–2556. https://doi.org/10.1111/gcb.16644.
(81) DeWitte, S. The Anthropology of Plague: Insights from Bioarcheological Analyses of Epidemic Cemeteries. *Mediev. Globe* **2014**, *1* (1), 97–124.
(82) Turchin, P. Modeling Social Pressures Toward Political Instability. *Cliodynamics* **2013**, *4* (2). https://doi.org/10.21237/C7clio4221333.
(83) Mieth, A.; Bork, H.-R. Humans, Climate or Introduced Rats – Which Is to Blame for the Woodland Destruction on Prehistoric Rapa Nui (Easter Island)? *J. Archaeol. Sci.* **2010**, *37* (2), 417–426. https://doi.org/10.1016/j.jas.2009.10.006.
(84) Moreno-Mayar, J. V.; Rasmussen, S.; Seguin-Orlando, A.; Rasmussen, M.; Liang, M.; Flåm, S. T.; Lie, B. A.; Gilfillan, G. D.; Nielsen, R.; Thorsby, E.; Willerslev, E.; Malaspinas, A.-S. Genome-Wide Ancestry Patterns in Rapanui Suggest Pre-European Admixture with Native Americans. *Curr. Biol.* **2014**, *24* (21), 2518–2525. https://doi.org/10.1016/j.cub.2014.09.057.
(85) Finney, B. Polynesian-South America Round Trip Canoe Voyages. **1994**.

(86) Archive, H. *Reefs and Atolls Around Mangareva and Rapa Nui*. https://archive.hokulea.com/rapanui/reefs.html (accessed 2025-02-15).
(87) SYSTEMS SCIENCE AND CHINA'S ECONOMIC REFORMS. In *Control Science and Technology for Development*; Pergamon, 1986; pp 1–7. https://doi.org/10.1016/B978-0-08-033473-8.50006-1.
(88) Chen, T.; Hou, P.; Wang, B.; Yang, J. Analysis of Fertility Policy Differences Based on Policy Continuity and Policy Effectiveness in China. *Biodemography Soc. Biol.* **2025**, *70* (3), 138–164. https://doi.org/10.1080/19485565.2025.2527033.
(89) Bongaarts, J.; Greenhalgh, S. An Alternative to the One-Child Policy in China. *Popul. Dev. Rev.* **1985**, *11* (4), 585–617.
(90) Bardi, U. *Pixie Dust and Overpopulation: What are the Origins of the "One Child" China's Population Policy?* The Seneca Effect. https://senecaeffect.substack.com/p/pixie-dust-and-overpopulation-what (accessed 2025-11-25).
(91) Greenhalgh, S. Missile Science, Population Science: The Origins of China's One-Child Policy. *China Q.* **2005**, No. 182, 253–276.
(92) Wang, T.; Jiang, Q. Recent Trend and Correlates of Induced Abortion in China: Evidence from the 2017 China Fertility Survey. *BMC Womens Health* **2022**, *22* (1), 469. https://doi.org/10.1186/s12905-022-02074-5.
(93) Pino, D. *China's Population Policy: A Failure and a Test*. National Review. https://www.nationalreview.com/corner/chinas-population-policy-a-failure-and-a-test/ (accessed 2025-11-28).
(94) Bardi, U.; El Asmar, T.; Lavacchi, A. Turning Electricity into Food: The Role of Renewable Energy in the Future of Agriculture. *J. Clean. Prod.* **2013**, *53*, 224–231. https://doi.org/10.1016/j.jclepro.2013.04.014.
(95) Maestas, N.; Mullen, K. J.; Powell, D. The Effect of Population Aging on Economic Growth, the Labor Force, and Productivity. *Am. Econ. J. Macroecon.* **2023**, *15* (2), 306–332. https://doi.org/10.1257/mac.20190196.
(96) Anu Madgavacar. *Confronting the Consequences of a New Demographic Reality | McKinsey*; McKinsey Institute, 2025. https://www.mckinsey.com/mgi/our-research/dependency-and-depopulation-confronting-the-consequences-of-a-new-demographic-reality (accessed 2025-03-24).

(97) Brijoux, T.; Neise, M.; Zank, S. Elder Abuse in the Oldest Old: Prevalence, Risk Factors and Consequences. *Z. Für Gerontol. Geriatr.* **2021**, *54* (2), 132–137. https://doi.org/10.1007/s00391-021-01945-0.

(98) Banerjee, A.; Yakovenko, V. Universal Patterns of Inequality. *New J. Phys.* **2010**, *12* (7), 075032. https://doi.org/10.1088/1367-2630/12/7/075032.

(99) Krawczyk, F.; Braun, A. Ch. Models like Heroes? Making Integrated Assessment Models (IAMs) Ready for Deep Decarbonization and a Socio-Economic Transformation. *Energy Res. Soc. Sci.* **2025**, *121*, 103959. https://doi.org/10.1016/j.erss.2025.103959.

(100) Solé, J.; Samsó, R.; García-Ladona, E.; García-Olivares, A.; Ballabrera-Poy, J.; Madurell, T.; Turiel, A.; Osychenko, O.; Álvarez, D.; Bardi, U.; Baumann, M.; Buchmann, K.; Capellán-Pérez, Í.; Černý, M.; Carpintero, Ó.; De Blas, I.; De Castro, C.; De Lathouwer, J.-D.; Duce, C.; Eggler, L.; Enríquez, J. M.; Falsini, S.; Feng, K.; Ferreras, N.; Frechoso, F.; Hubacek, K.; Jones, A.; Kaclíková, R.; Kerschner, C.; Kimmich, C.; Lobejón, L. F.; Lomas, P. L.; Martelloni, G.; Mediavilla, M.; Miguel, L. J.; Natalini, D.; Nieto, J.; Nikolaev, A.; Parrado, G.; Papagianni, S.; Perissi, I.; Ploiner, C.; Radulov, L.; Rodrigo, P.; Sun, L.; Theofilidi, M. Modelling the Renewable Transition: Scenarios and Pathways for a Decarbonized Future Using *Pymedeas*, a New Open-Source Energy Systems Model. *Renew. Sustain. Energy Rev.* **2020**, *132*, 110105. https://doi.org/10.1016/j.rser.2020.110105.

(101) Nordhaus, W. *The "DICE" Model: Background and Structure of a Dynamic Integrated Climate-Economy Model of the Economics of Global Warming*; Cowles Foundation Discussion Papers; Working paper 1009; Cowles Foundation for Research in Economics, Yale University, 1992. https://EconPapers.repec.org/RePEc:cwl:cwldpp:1009 (accessed 2025-08-06).

(102) Bardi, U.; Bierwirth, P.; Huang, K.-W.; MacIntyre, J. Carbon Dioxide as a Pollutant. The Risks on Human Health and the Stability of the Biosphere. *Environ. Sci. Adv.* **2025**. https://doi.org/10.1039/D5VA00017C.

(103) Duncan, R.C. The Olduvai Theory, Sliding Towards a Post-Industrial Stone Age. *Soc. Contract* **2005**, *16* (2), 1–12.

(104) Klein, S. *Infographics in the Time of Cholera*. ProPublica. https://www.propublica.org/nerds/infographics-in-the-time-of-cholera (accessed 2025-03-13).

(105) Gannett, H. *Statistical Atlas of the United States, Based upon the Results of the Eleventh Census*; United States. Census office. 11th census, 1890., 1893.

(106) Kermack, W. O.; McKendrick, A. G.; Walker, G. T. A Contribution to the Mathematical Theory of Epidemics. *Proc. R. Soc. Lond. Ser. Contain. Pap. Math. Phys. Character* **1927**, *115* (772), 700–721. https://doi.org/10.1098/rspa.1927.0118.

(107) Perissi, I.; Lavacchi, A.; Bardi, U. Peaking Dynamics of the Production Cycle of a Nonrenewable Resource. *Sustainability* **2023**, *15* (8), 1–11.

(108) Taleb, N. *The Black Swan*; Random House, 2007.

(109) Sornette, D. Dragon-Kings, Black Swans and the Prediction of Crises. *ArXiv09074290v1 Physicsdata-An* **2009**, 18.

(110) Bak, P. *How Nature Works. The Science of Self-Organized Criticality.*; Copernicus: New York, 1996.

(111) Turner, G. A Comparison of The Limits to Growth with 30 Years of Reality. *Glob. Environ. Change* **2008**, *18* (3), 397–411. https://doi.org/10.1016/j.gloenvcha.2008.05.001.

(112) Herrington, G. Update to Limits to Growth: Comparing the World3 Model with Empirical Data. *J. Ind. Ecol.* **2021**, *25* (3), 614–626. https://doi.org/10.1111/jiec.13084.

Printed in Dunstable, United Kingdom

84347386R00171